Merchant Ship
Stability

Marine Engineering Series

MARINE AUXILIARY MACHINERY—6th edition
David W Smith, CEng, MIMarE

POUNDER'S MARINE DIESEL ENGINES—6th edition
C T Wilbur, CEng, MIMarE and
D A Wight, BSc, CEng, MIMechE, FIMarE

MARINE ELECTRICAL PRACTICE—5th edition
G O Watson, FIEE, FAIEE, FIMarE

MARINE AND OFFSHORE CORROSION
Kenneth A Chandler, BSc, CEng, FIM, ARSM, FICorrT

MARINE AND OFFSHORE PUMPING AND PIPING SYSTEMS
J Crawford, CEng, FMarE

MARINE STEAM BOILERS—4th edition
J H Milton, CEng, FIMarE, and
Roy M Leach, CEng, MIMechE, FIMarE

MARINE STEAM ENGINES AND TURBINES—4th edition
S C McBirnie, CEng, FIMechE

MERCHANT SHIP STABILITY
Captain A R Lester, Extra Master, BA(Hons), MRINA, MNI

Merchant Ship Stability

Captain A R Lester, Extra Master, BA(Hons), MRINA, MNI
Faculty of Maritime Studies, Plymouth Polytechnic

BUTTERWORTHS
London − Boston − Durban − Singapore − Sydney − Toronto − Wellington

First published 1985

© Butterworth & Co (Publishers) Ltd, 1985

British Library Cataloguing in Publication Data

Lester, A.R.
    Merchant ship stability.—(Marine engineering
series)
    1. Merchant ships—Stability
    I. Title   II. Series
    623.8'171   VM159

    ISBN 0–408–01448–2

Library of Congress Cataloging in Publication Data

Lester, A. R. (Alan Robert)
    Merchant ship stability.

    Bibliography: p.
    Includes index.
    1. Stability of ships.   2. Merchant ships.
I. Title.
VM159.L47 1985   623.8'171   85–6621
ISBN 0–408–01448–2

Filmset by Mid-County Press, 2a Merivale Road, London SW15 2NW
Printed and bound in England by Garden City Press Ltd., Letchworth, Herts

Preface

This book is intended to serve the needs of the deck officer in a merchant ship, as well as first year degree students and others at a similar level. The work covered is sufficient to deal with the ship stability and stress syllabus from cadet entry to master. Whenever possible the practical application of theory to ship operations is presented in a form which is currently being applied in practice.

It is assumed that readers will have a basic knowledge of physics sufficient to deal with ideas of moments and elementary hydrostatics. Wherever possible, alternatives to the methods of calculus have been used; however, an elementary knowledge of integration and differentiation will enable the reader to gain a full understanding of basic principles

Chapter 1 contains basic work on the form and measurement of ships. The methods of Simpson's Rules have been used to present a means of measuring ship form. While there are much more efficient methods available, especially for computer application, Simpson's Rules are much easier to present and to manipulate using calculators. Chapter 1 should be used for reference as necessary rather than read straight through.

Chapters 2 to 10 are progressive as all information needed to deal with current work is contained in previous chapters. As far as possible the work presented within each chapter is also progressive, so that it is possible for the elementary student to have a course of study based upon the early parts of each chapter. The more advanced student can make use of more information while having revision material at hand.

Chapter 11 'Stress' and Chapter 12 'Ship behaviour' both have links with the previous work but are largely independent of the early chapters.

Specific examples of computer applications are not given. However, where a particular process is generally applied as a computer program, this is indicated in the text.

The author and publishers would like to thank Denholm Ship Management Limited for permission to use the hydrostatic, stability and stress data for one of their bulk carriers as a basis for the *MV Nonesuch* data which is used throughout the book.

A R Lester

Author's Note

Symbols are defined as they arise in the text, the following general conventions have been used:

Upper case letters represent total values, i.e. W for displacement, B for buoyancy, KG for distance of centre of gravity of the ship above the keel.

Lower case letters represent elements of the total, i.e. w for a component weight, Kg for the distance of a component weight from the keel.

Where there is movement of centres of gravity, buoyancy, etc. then the suffix o is used to indicate the original position and subsequent positions by suffixes 1, 2, etc. In diagrams dashed lines are used to represent initial conditions and solid lines final conditions.

Acknowledgements

I would like to thank my father John Lester for checking the arithmetic and Jeffrey Cornwall and Richard Dakin for checking examples in the later chapters. Any remaining errors are entirely my fault.

The author and publishers would also like to thank Mr P W Penney of the Department of Naval Architecture and Shipbuilding, The University of Newcastle upon Tyne for his helpful suggestions in connection with the jacket design.

Contents

1 Ship form and measurement

This chapter deals with the techniques of measurement which need to be used in order to compute the various quantities which enable hydrostatic and stability data for ships to be determined. It is not intended to be read before the following chapters but is intended as reference material. As such it should be consulted when particular topics are reached in the subsequent work, where the techniques contained in this chapter are required.

OBJECTIVES

1. To define the dimensions and form of ships.
2. To compute areas using the trapezoidal rule; Simpson's first, second and 5, 8, −1 Rules.
3. To apply the techniques in (2) to finding the first moment of areas and volumes.
4. To apply the techniques in (2) to finding the second moments of areas.
5. To find the positions of the centroids of areas and volumes.
6. To find the second moments of areas about an axis through the centroid.
7. To apply the values calculated to the determination of the hydrostatic and stability data of ships and other floating platforms.

DEFINITIONS

See *Figures 1.1* and *1.2*.

Length between perpendiculars (L_{pp})
The distance in metres on the summer load waterline from the fore side of

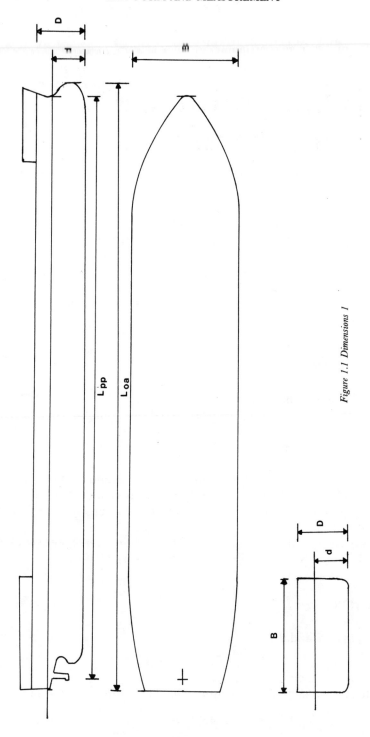

Figure 1.1 Dimensions 1

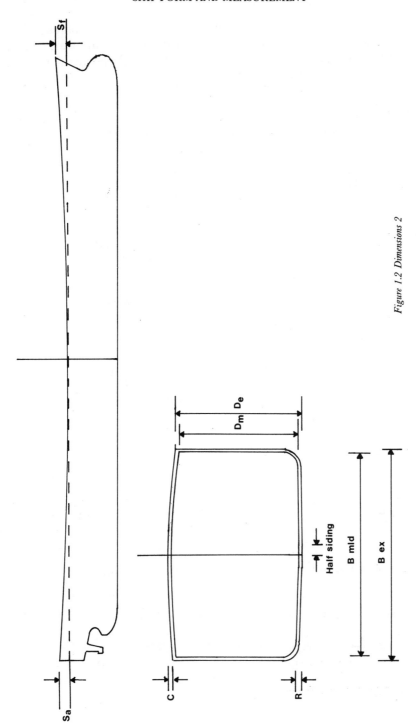

Figure 1.2 Dimensions 2

the stem to the after side of the rudder post; or to the centre of the rudder stock if there is no rudder post. These forward and aft positions are termed the 'Forward Perpendicular' (FP) and 'After Perpendicular' (AP) respectively.

Length overall (L_{oa})
The maximum length of the vessel from bow to stern.

Breadth (*B*)
The greatest breadth of the vessel.

Depth (*D*)
The depth of the vessel from the deck to the keel amidships.

Draft (*d*) (In some references, denoted '*T*')
The distance from the waterline to the keel amidships.

Breadth, depth and draft may be either extreme, when the measurements are taken to the outside of the shell plating, or moulded, when they are taken to the inside of the shell plating. References to a dimension in this book are to its extreme value, outside the shell plating, unless otherwise stated.

Sheer
The rise in the deck line from amidships to the forward and after ends of the vessel.

Camber
The rise of the deck from the side of the vessel to the centreline.

Rise of floor
The rise of the ship's bottom from the keel to the bilge.

RELATIONSHIPS BETWEEN DIMENSIONS

In order to ensure that conventional ship shapes have reasonable stability and stress characteristics, constraints are imposed on the relationships between the principle dimensions of ships. These relationships are expressed as the ratio one dimension bears to another.

For typical ship shapes, the range of values which will normally be found are as follows:

$$\frac{L}{D} = 10\text{--}6$$

$$\frac{L}{B} = 5.5\text{--}8$$

$$\frac{B}{d} = 2\text{--}3$$

MEASUREMENT OF AREA

Any quantity which can be represented by the area under a curve can be determined by measuring the area under the curve between limits. If the equation of the curve is known, then the techniques of integral calculus can be used. If, as is generally the case with ship data, there is no equation which describes the bounding curve, one of the techniques of numerical integration must be used.

All of these techniques rely upon dividing the area to be measured into segments and then applying various formulae to the ordinates bounding the segments (*Figure 1.3*).

In general:

$$\text{Area} = C(K_0 y_0 + K_1 y_1 + \cdots K_6 y_6)$$

where C, K_0, K_1 etc. are constant multipliers by which the ordinate values y_0, y_1 etc. are multiplied.

If the bounding curve is now assumed to be a polynomial of some

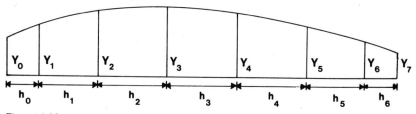

Figure 1.3 Measurement of Area

order. (Note that a polynomial is any equation which measures the value of y by saying that:

$$y = ax^n + bx^{n-1} + \cdots px + q)$$

there are three alternative approaches to determining the area under the curve using numerical integration:

(a) To assume that the intervals between the ordinates, $h_0, h_1 \ldots h_5$ are all equal and to find the multipliers $C, K_0, K_1 \ldots K_6$ by which the ordinate values $y_0, y_1 \ldots y_6$ are to be multiplied in order to give the area. This approach is used to produce Simpson's Rules, which will be used in this book to determine areas.

(b) To allow the intervals $h_0, h_1 \ldots h_5$ to vary and then to find the intervals required to make the multipliers $K_0 = K_1 = \ldots K_6 = 1$. This approach produces Tchybecheff's rules, which will not be examined further in this book.

(c) To allow both the intervals $h_0, h_1 \ldots h_5$ and the multipliers $K_0, K_1 \ldots K_6$ to vary. This approach produces Gauss rules, which will not be examined further in this book.

There is an improvement in efficiency from Method (a) to Method (c); i.e. fewer computations have to be performed in order to achieve a given degree of accuracy. Simpson's Rules have the advantage that the formulae produced by them are easily manipulated using a hand calculator. Methods (b) and (c) are more suited to computer calculations and may be studied further in Reference 1.

Simpson's Rules

(i) *The Trapezoid Rule (Figure 1.4)*
This rule assumes that the bounding curve between two ordinates y_0 and y_1 is a straight line, i.e.

$$y = ax + c$$

Hence: Area $= \dfrac{h}{2}(y_0 + y_1)$

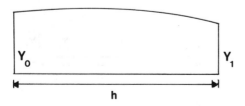

Figure 1.4 Trapezoidal Rule

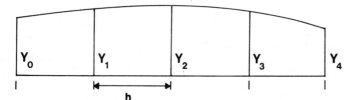

Figure 1.5 Compound Trapezoidal Rule

The rule may be extended to any number of ordinates and is then known as the Compound Trapezoid Rule, in which (*Figure 1.5*):

$$\text{Area} = \frac{h}{2}\left(y_0 + 2y_1 + 2y_2 + 2y_3 + y_4\right)$$

This rule gives area values which are less than those existing under a curve, but may usefully be applied as a check on areas found using more complex rules.

(ii) *Simpson's First Rule (Figure 1.6)*
This rule assumes that the bounding curve, defined by three ordinates y_0, y_1 and y_2, is a second order polynomial, i.e.

$$y = ax^2 + bx + c$$

From which the area is found to be:

$$\text{Area} = \frac{h}{3}\left(y_0 + 4y_1 + y_2\right)$$

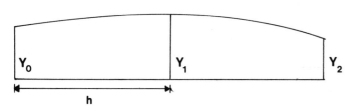

Figure 1.6 Simpson's First Rule

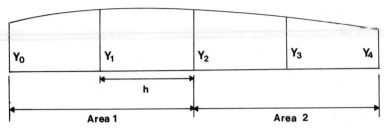

Figure 1.7 Simpson's compound First Rule

This rule may be extended to any area wjich is divided by an **odd** number of ordinates, as follows:

The Compound First Rule obtains the total area under the curve by adding together the areas defined by groups of three ordinates, as follows (*Figure 1.7*):

$$\text{Total Area} = \text{area } 1 + \text{area } 2$$

i.e. $\text{Total Area} = \dfrac{h}{3}(y_0 + 4y_1 + y_2) + \dfrac{h}{3}(y_2 + 4y_3 + y_4)$

$$= \dfrac{h}{3}(y_0 + 4y_1 + 2y_2 + 4y_3 + y_4)$$

In a similar way, any area with an **odd** number of ordinates can be found using the first rule. The multipliers inside the brackets are 1, 4, 2, 4, 2, 4, 2, 4, ..., 1.

A derivation of the first rule is now presented to illustrate the general principles governing the derivation of Simpson's Rules (*Figure 1.8*).

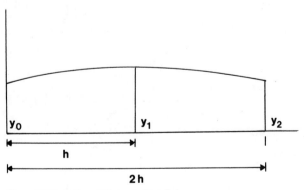

Figure 1.8 Derivation of Simpson's First Rule

If $y = ax^2 + bx + c$

$$\text{Area} = \int_0^{2h} (ax^2 + bx + c)$$

$$= \left[\frac{ax^3}{3} + \frac{bx^2}{2} + cx + d \right]_0^{2h}$$

$$= \frac{8h^3}{3} a + \frac{4h^2}{2} b + 2hc$$

Now when:

$$x = 0$$

$$y_0 = c \qquad\qquad\qquad\qquad\text{(i)}$$

When
$$x = h$$

$$y_1 = h^2 a + hb + c \qquad\qquad\qquad\qquad\text{(ii)}$$

When
$$x = 2h$$

$$y_2 = 4h^2 + 2hb + c \qquad\qquad\qquad\qquad\text{(iii)}$$

$$y_1 - y_0 = h^2 a + hb$$

$$y_2 - y_0 = 4h^2 a + 2hb$$

$$(y_2 - y_0) - 2(y_1 - y_0) = 2h^2 a$$

$$\therefore \quad a = \frac{y_0 - 2y_1 + y_2}{2h^2}$$

$$(y_2 - y_0) - 4(y_1 - y_0) = -2hb$$

$$\therefore \quad b = \frac{-3y_0 + 4y_1 - y_2}{2h}$$

$$c = y_0$$

Substituting in the formula for the area:

$$\text{Area} = \frac{8h^3}{3}\frac{(y_0 - 2y_1 + y_2)}{2h^2} + \frac{4h^2}{2}\frac{(-3y_0 + 4y_1 - y_2)}{2h} + 2h \cdot y_0$$

$$= \frac{4h}{3}(y_0 - 2y_1 + y_2) + h(-3y_0 + 4y_1 - y_2) + 2hy_0$$

$$= h\left(\frac{4}{3}y_0 - \frac{8}{3}y_1 + \frac{4}{3}y_2 - 3y_0 + 4y_1 - y_2 + 2y_0\right)$$

$$= h\left(\frac{1}{3}y_0 + \frac{4}{3}y_1 + \frac{1}{3}y_2\right)$$

i.e.　$\text{Area} = \dfrac{h}{3}(y_0 + 4y_1 + y_2)$

The First Rule can be further extended to reduce computation when some segments of an area are bounded by a curve which is changing rapidly, while other parts of the curve are changing very slowly or not at all. For example, the bow and stern portions of a ship's shape are parts of the ship where shape changes rapidly, while over the parallel middle body the shape remains constant.

The First Rule is modified by the insertion of intermediate ordinates which are measured at points half way between two of the main ordinates, as follows (*Figure 1.9*):

$$\text{Area } 1 = \frac{1}{3}\frac{h}{2}(y_0 + 4y_1 + y_2) = \frac{1}{3}h\left(\frac{y_0}{2} + 2y_1 + \frac{y_2}{2}\right)$$

$$\text{Area } 2 = \frac{1}{3}h(y_2 + 4y_3 + y_4)$$

$$\text{Area } 3 = \frac{1}{3}\frac{h}{2}(y_4 + 4y_5 + y_6) = \frac{1}{3}h\left(\frac{y_4}{2} + 2y_5 + \frac{y_6}{2}\right)$$

$$\text{Total Area} = \frac{1}{3}h\left(\frac{y_0}{2} + 2y_1 + \frac{y_2}{2}\right) + \frac{1}{3}h(y_2 + 4y_3 + y_4)$$

$$+ \frac{1}{3}h\left(\frac{y_4}{2} + 2y_5 + \frac{y_6}{2}\right)$$

$$\text{Total Area} = \frac{h}{3}\left(\frac{y_0}{2} + 2y_1 + \frac{3}{2}y_2 + 4y_3 + \frac{3}{2}y_4 + 2y_5 + \frac{y_6}{2}\right)$$

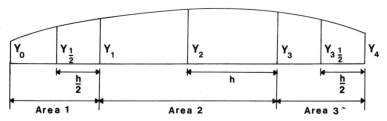

Figure 1.9 Use of intermediate ordinates

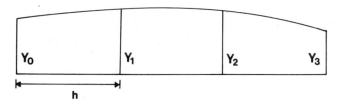

Figure 1.10 Simpson's Second Rule

(iii) *Simpson's Second Rule (Figure 1.10)*
This rule assumes that the bounding curve, defined by four ordinates y_0, y_1, y_2 and y_3 is a third order polynomial, i.e.

$$y = ax^3 + bx^2 + cs + d$$

From which the area is found to be:

$$\text{Area} = \frac{3}{8} h(y_0 + 3y_1 + 3y_2 + y_3)$$

This rule may be extended to any area which is defined by a number of ordinates such that the number of spaces between the ordinates can be divided by three; i.e. it is applied to 4, 7, 10, 13 ... ordinates, the number of spaces being 3, 6, 9, 12 ...

The Compound Second Rule obtains the total area under the curve by adding together the areas defined by groups of four ordinates, as follows (*Figure 1.11*):

Total Area = Area 1 + Area 2

i.e.   Total Area $= \dfrac{8}{3} h(y_0 + 3y_1 + 3y_2 + y_3) + \dfrac{3}{8} h(y_3 + 3y_4 + 3y_5 + y_6)$

∴   Total Area $= \dfrac{3}{8} h(y_0 + 3y_1 + 3y_2 + 2y_3 + 3y_4 + 3y_5 + y_6)$

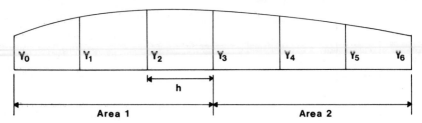

Figure 1.11 Simpson's compound Second Rule

The multipliers inside the brackets for the Second Rule are 1, 3,3, 2, 3, 3, 2, 3, 3, 2,..., 1. This rule is therefore used to obtain areas when the First Rule will not fit. Where both the First and the Second Rules will fit, it is conventional to use the First Rule in preference to the Second.

Where neither of these rules fit, then the area must be divided into subsections, each area found separately and the sum of the areas found.

A Simpson's Rule can be found for any number of ordinates. For example, the rule for six ordinates may sometimes be required and is now stated for reference purposes:

$$\text{Area} = \frac{25}{24} h\left(\frac{2}{5}y_0 + y_1 + y_2 + y_3 + y_4 + y_5 + \frac{2}{3}y_6\right)$$

The 5, 8, −1 Rule:

A derivative of the First Rule may be used to find the area between any two successive ordinates when three ordinates defining a curve are given. This rule is called the 5, 8, −1 Rule because of the multipliers used inside the brackets. If an area defined by three ordinates is defined, then it may be shown that (*Figure 1.12*):

$$\text{Area } 1 = \frac{h}{12}(5y_0 + 8y_1 - y_2)$$

$$\text{Area } 2 = \frac{h}{12}(5y_2 + 8y_1 - y_0)$$

∴ $$\text{Area } 1 + \text{Area } 2 = \frac{h}{12}(4y_0 + 16y_1 + 4y_2)$$

i.e. $$\text{Area } 1 + \text{Area } 2 = \frac{h}{3}(y_0 + 4y_1 + y_2)$$

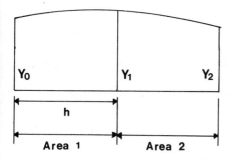

Figure 1.12 Simpson's 5 8-1 rule

Appendages

It is often the case that there are small areas which it is inconvenient to include within the main area to be measured using the rules. These appendages must be measured separately and added to the main area. In many cases these appendages will be approximately triangular in shape.

EXAMPLES

The worked examples which follow illustrate the method of obtaining areas under curves using Simpson's Rules. The ordinates are measured at a number of stations along the length of the area. For most problems relating to ship shapes, the waterplane is involved and stations are measured along the centreline of the vessel. The ordinates are then measured to the shell plating of the ship on one side only, it being assumed that port and starboard sides are identical. Ordinates measured on one side only of the centreline are termed half ordinates.

When solving problems involving the use of Simpson's Rules, the following points should be carefully remembered:

1. Establish how many ordinates (stations) define the area and draw a sketch to illustrate the area to be found.
2. Determine which rule or combination of rules should be used to obtain the area.
3. Check if ordinates are equally spaced or not. Particular care in determining the multipliers must be used in problems involving the use of intermediate ordinates and the First Rule.
4. Lay the work out neatly to minimize the risk of mistakes and to enable checking to be carried out easily.
5. Ensure that the answer obtained is a reasonable one by making a check using the Trapezoid Rule or mentally obtaining the 'mean' ordinate and multiplying this by the length of the area to obtain an approximation to the area.

If the problem to be solved is not one set in a book or examination, i.e. where the reader has to go out and measure, say, the curved area of a tank top in a ship, then the selection of station spacing to make the calculation of the area using either the First or Second Rules as convenient as possible is clearly advisable.

Example 1.1

A vessel of length 90 m has equally spaced half ordinates of the waterplane as follows, commencing from the after perpendicular:

Station	0	1	2	3	4	5	6
½ Ordinate (m)	0.1	2.4	2.7	2.8	2.8	2.2	0.2

Find the area of the waterplane.

Station	½ Ord (m)	Simpson's multiplier	Function for area
0	0.1	1	0.1
1	2.4	4	9.6
2	2.7	2	5.4
3	2.8	4	11.2
4	2.8	2	5.6
5	2.2	4	8.8
6	0.2	1	0.2
			Σ 40.9

$$\text{Area} = 2 \times \frac{1}{3} \times h\,\Sigma\,\text{Function for Area}$$

$$= 2 \times \frac{1}{3} \times 15 \times 40.9$$

$$= 409 \text{ m}^2$$

The value of 'h' is found by dividing the length of the waterplane by the number of spaces, i.e.

90 m divided by 6 spaces = 15 m.

Example 1.2

A vessel of length 300 m has half ordinates of waterplane as follows, commencing from the after perpendicular:

Station	0	$\frac{1}{2}$	1	2	3	4	5	$5\frac{1}{2}$	6
$\frac{1}{2}$ Ordinate (m)	0.1	7.5	10	12	12.3	11.4	8	5.2	1.0

Find the area of the waterplane if there are appendages forward and aft with a total area of 2.8 m^2 to be added to the main area.

Station	$\frac{1}{2}$ Ord (m)	Simpson's multiplier	Function for area
0	0.1	$\frac{1}{2}$	0
$\frac{1}{2}$	7.5	2	15.0
1	10.0	$1\frac{1}{2}$	15.0
2	12.0	4	48.0
3	12.3	2	26.6
4	11.4	4	45.6
5	8.0	$1\frac{1}{2}$	12.0
$5\frac{1}{2}$	5.2	2	10.4
6	1.0	$\frac{1}{2}$	0.5
			Σ 173.1

$$\text{Area} = \left[2 \times \frac{h}{3} \times \Sigma \text{ Function of Area} \right] + \text{Appendage Area}$$

$$= \left(2 \times \frac{50}{3} \times 173.1 + 2.8 \right) \text{m}^2$$

$$= 5772.8 \text{ m}^2$$

In the example above, there are 7 stations (0, 1, 2, 3, 4, 5 and 6) and two intermediate stations, $\frac{1}{2}$ and $5\frac{1}{2}$. The station spacing is therefore

$$\frac{\text{length of vessel}}{\text{number of spaces}} = \frac{300}{6} = 50 \text{ m}.$$

Example 1.3

A vessel has the following waterplane areas at the drafts given:

Draft (m)	0	2	4	6	8	10
Area (m²)	350	2500	3450	3960	4000	4030

Find displacement of vessel in salt water at drafts of 2 m, 6 m, and 10 m.

At 2 m draft

Station (draft)	Area (m²)	Simpson's multiplier	Function for volume
0	350	5	1750
2	2500	8	20 000
4	3450	−1	−3450
			Σ 18 300

$$\text{Volume} = \frac{h}{12} \times \Sigma \text{ Function of volume}$$

$$= \frac{2}{12} \times 18\,300 = 3050 \text{ m}^3$$

Displacement = Volume × density
= 3050 × 1.025 = 3126.25 tonnes

At 6 m draft

Station (draft)	Area (m²)	Simpson's multiplier	Function for volume
0	350	1	350
2	2500	3	7500
4	3450	3	10350
6	3960	1	3960
			Σ 22160

$$\text{Volume} = \frac{3}{8} \times h \times \Sigma \text{ Function of volume}$$

$$= \frac{3}{8} \times 2 \times 22\,160 = 16\,620 \text{ m}^3$$

Displacement = 16 620 × 1.025 = 17 035.5 tonnes

At 10 m draft

Station (draft)	Area (m²)	Simpson's multiplier 0–4	Function for volume 0–4	Simpson's multiplier 4–10	Function for volume 4–10
0	350	1	350		
2	2500	4	10000		
4	3450	1	3450	1	3450
6	3960			3	11880
8	4000			3	12000
10	4030			1	4030
			Σ 13800		Σ 31360

$$\text{Volume}_{0-4} = \frac{h}{3} \times \Sigma \text{ Function of Volume}$$

$$= \frac{2}{3} \times 13\,800 = 9200 \text{ m}^3$$

$$\text{Volume}_{4-10} = \frac{3}{8} \times h \times \Sigma \text{ Function of Volume}$$

$$= \frac{3}{8} \times 2 \times 31\,360 = 23\,520 \text{ m}^3$$

$\therefore$ Total Volume$_{0-10} = 8200 + 23\,520 = 31\,720$ m^3

Displacement $= 31\,720 \times 1.025 = 32\,513$ tonnes

By using the six ordinate rule, the volume$_{0-10} = 32\,629.2$ m^3.

MOMENTS OF AREAS AND VOLUMES AND CENTROIDS

First moments of areas and volumes

The value of the first moment of an area is required in order to calculate the position of the centroid of the area. The position of the centroid represents the position of the centre of floatation of a waterplane and the position of the centre of buoyancy of the underwater volume of a vessel.

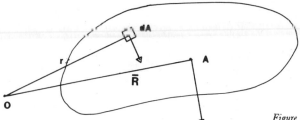

Figure 1.13 First moment of Area

Consider an area A with an element of area δA at distance r from an axis passing through 0 (*Figure 1.13*)

First moment of the element $\delta A = r\,\delta A$

First moment of full area $\quad = \Sigma\, r\,\delta A$

First moment of full area $\quad = \bar{R}A$

$$\bar{R}A = \Sigma\, r\,\delta A$$

$$\bar{R} = \frac{\Sigma\, r\,\delta A}{A}$$

i.e.

$$\bar{R} = \frac{\text{First moment of area}}{\text{area}}$$

In a similar way

$$\bar{R} = \frac{\text{First moment of volume}}{\text{volume}}$$

Second moments of areas (*Figure 1.13*)

The second moment of area is a mathematical form which arises from the moment of inertia of a rotating mass. In general, the energy of a body can be written as:

$$\text{Total Energy} = \text{Potential Energy} + \text{Kinetic Energy}$$
$$+ \text{Rotational Energy}$$
$$= mgh + \tfrac{1}{2}mv^2 + \tfrac{1}{2}I\omega^2$$

where $\qquad\omega =$ angular velocity;

$I =$ the moment of inertia about the axis of rotation;
$= Cma^2$ (where $C =$ a constant; $m =$ the mass of the body; $a =$ some dimension of the body).

Similar mathematics to the above can be used to find the second moment of volume or the second moment of area. The term second moment is used because of the a^2 term in the expression. Cma is the first moment, used to find the centre of gravity of a body, Cma^2 is the second moment.

It is not surprising that inertia should play some part in assessing the stability of ships and other floating bodies, because ships are rotating about some axis as they roll and pitch. The mathematical form of second moments appear when analysing transverse and longitudinal stability.

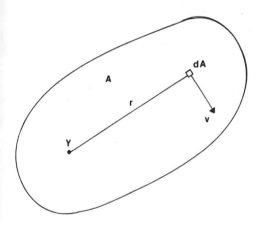

Figure 1.14 Second moment of Area

Consider an area A rotating about some axis Y with an angular velocity w_1 (*Figure 1.14*).

A is an element of the area at a distance r from the axis Y.

$$\text{The Kinetic Energy of the Element} = \tfrac{1}{2}\delta A v^2$$

$$V = rw$$

$\therefore$ $$\text{Kinetic Energy (KE)} = \tfrac{1}{2}\delta A r^2 w^2$$

and for the full area A:

$$KE = \Sigma \tfrac{1}{2}\delta A r^2 w^2$$
$$= \tfrac{1}{2}w^2 \Sigma \delta A r^2$$

The value of $\Sigma \delta A r^2$ will be a constant when considering the rotation of the area about any given axis.

Ar^2 is the second moment of the area about the axis Y.

We require two values of second moments for our work on ship stability and will shortly investigate one of the several theorems which connect the second moments of an area about different axes.

Routine work on rectangles makes use of the values of second moment about the centroid of the rectangle and also about its edge, as follows:

Second moment of a rectangle about an axis through its centroid (Figure 1.15):

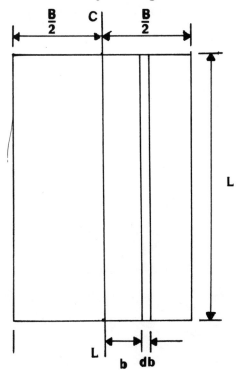

Figure 1.15 *Second moment of rectangle about centreline*

$$\text{Second moment of a strip} = Ldbb^2$$

$$Lb^2\,db$$

$$\therefore \quad \text{Second moment of full area} = \int_{-B/2}^{B/2} Lb^2\,db$$

$$= 2\int_{0}^{B/2} Lb^2\,db$$

$$= 2\left[\frac{Lb^3}{3}\right]_0^{B/2}$$

$$= 2L\frac{B^3}{8}\times\frac{1}{3}=\frac{LB^3}{12}$$

Second moment of a rectangle about one side (Figure 1.16)

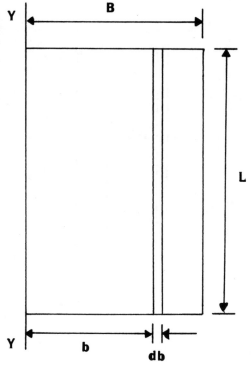

Figure 1.16 *Second moment of rectangle about edge*

$$\text{Second moment of a strip} = \int_0^B Lb^2 \, db$$

$$= \left[\frac{Lb^3}{3} \right]_0^B = \frac{LB^3}{3}$$

The parallel axes theorem

The parallel axes theorem is used in cases where we wish to determine the second moment of an area about one axis (**YY**) and we already know or can find the value of the second moment of the area about another axis which is parallel to YY (*Figure 1.17*).

Consider the second moment of the area about axis YY, given that NA is an axis parallel to YY and that NA passes through the centroid of the area, located at C.

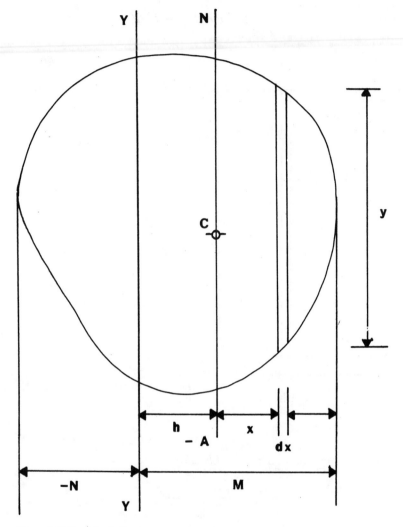

Figure 1.17 Parallel axis theorem

$$I_{YY} = \int_{-N}^{M} (h+x)^2 y \, dx$$

$$= \int_{-N}^{M} (h^2 + 2hx + x^2) y \, dx$$

$$= \int_{-N}^{M} h^2 y \, dx + \int_{-N}^{M} 2hxy \, dx + \int_{-N}^{M} x^2 y \, dx$$

Now $\displaystyle\int_{-N}^{M} h^2 y\, dx = h^2 \int_{-N}^{M} y\, dx = h^2 A$

and

$$\int_{-N}^{M} 2hxy\, dx = 2h \int_{-N}^{M} xy\, dx = 0$$

(since $\displaystyle\int_{-N}^{M} xy\, dx$ is the first moment about the neutral axis)

Also $\displaystyle\int_{-N}^{M} x^2 y\, dx = I_{NA}$

$\therefore \qquad I_{YY} = I_{NA} + h^2 A$

or

$$I_{NA} = I_{YY} - h^2 A$$

i.e. the parallel axes theorem states that the second moment of an area about an axis (YY) which is parallel to an axis passing through the centroid of the area (NA) is the sum of the second moment about the axis through the centroid and the product of the area and the square of the distance between the axes.

Note that, when the parallel axes theorem is applied to a rectangle to determine the second moment about an edge, knowing that the second moment of the rectangle about the centreline is $LB^3/12$:

$$I_{edge} = I_{CL} + Ah^2$$

i.e. $\quad I_{edge} = \dfrac{LB^3}{12} + LB\left(\dfrac{B}{2}\right)^2$

$$= \dfrac{LB^3}{12} + \dfrac{LB^3}{4} = \dfrac{LB^3}{3}$$

which confirms the derivation of the second moments about the centreline and edge for a rectangle.

APPLICATION OF SIMPSON'S RULES

In working with waterplanes and underwater volumes of ships and other structures, we need to apply Simpson's Rules to finding:

Areas and volumes of curved shapes
First moments of these areas and volumes
Second moments of curved areas

The first moments of the areas or volumes are used to find the distance of the centroid of the area or volume from a given axis, while the second moments of areas are used in the calculation of initial transverse and longitudinal stability.

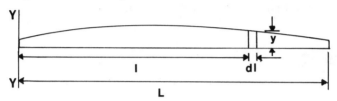

Figure 1.18 Area, second moment and second moment about YY

In *Figure 1.18*.

$$\text{Area} = \int_0^L y \, dl$$

$$\text{First moment of the area} = \int_0^L yl \, dl$$

relative to the transverse axis YY

$$\text{Second moment of the area} = \int_0^L yl^2 \, dl$$

relative to the transverse axis YY

The effect of translating these quantities into areas is shown in *Figure 1.19* where (a) is area, (b) is the first moment of the area and (c) is the second moment of the area. *Tables 1.1* to *1.4* give the standard proforma for calculating these values using Simpson's Rules.

Table 1.4 gives the standard proforma for calculating values relative to axis YY using Simpson's Rules. This work has been reduced by noting that for, say, y_2 in *Table 1.2*

$$F \text{ (first moments)} = F \text{ (area)} \times 2h = 4y_2$$

and in *Table 1.3*

$$F \text{ (second moments)} = F \text{ (first moments)} \times 2h = 8h^2 y_2$$

also that h and h^2 are common factors in the tables for first moments and second moments respectively.

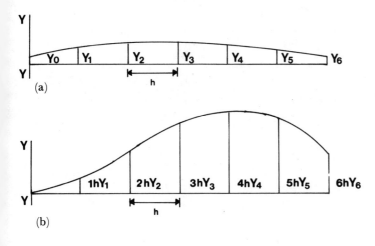

(a)

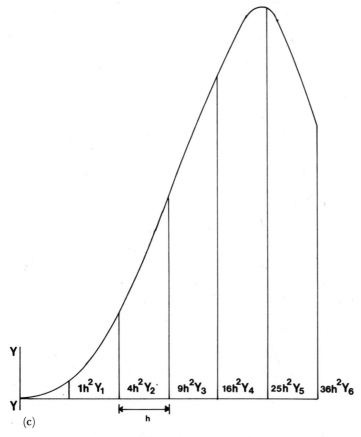

(b)

(c)

Figure 1.19 (a) Area (Figure 1.18). (b) First moment of Area of (a). (c) Second moment of Area of (a)

Table 1.1 Area

Station	Ord	Simpson's multiplier	F (Area)
0	y_0	1	y_0
1	y_1	4	$4y_1$
2	y_2	2	$2y_2$
3	y_3	4	$4y_3$
4	y_4	2	$2y_4$
5	y_5	4	$4y_5$
6	y_6	1	y_6

$$\Sigma F \text{ (Area)}$$

Table 1.2 First moments

Station	Ord	Simpson's multiplier	F (First moments)
0	$0y_0$	1	0
1	hy_1	4	$4hy_1$
2	$2hy_2$	2	$4hy_2$
3	$3hy_3$	4	$12hy_3$
4	$4hy_4$	2	$8hy_4$
5	$5hy_5$	4	$20hy_5$
6	$6hy_6$	1	$6hy_6$

$$\Sigma F \text{ (first moments)}$$

$$\text{and first moments} = \frac{h}{3} \Sigma F \text{ (first moments)}$$

Table 1.3 Second moments

Station	Ord	Simpson's multiplier	F (Second moments)
0	$0y_0$	1	0
1	h^2y_1	4	$4h^2y_1$
2	$4h^2y_2$	2	$8h^2y_2$
3	$9h^2y_3$	4	$36h^2y_3$
4	$16h^2y_4$	2	$32h^2y_4$
5	$25h^2y_5$	4	$100h^2y_5$
6	$36h^2y_6$	1	$36h^2y_6$

$$\Sigma F \text{ (second moments)}$$

$$\text{and second moments} = \frac{h}{3} \Sigma F \text{ (second moments)}$$

Table 1.4

Station	Ord	Simpson's multiplier	F (Area)	Lever First moment	F (First moment)	Lever (Second moment)	F (Second moment)
0	y_0	1	y_0	0	0	0	0
1	y_1	4	$4y_1$	1	$4y$	1	$4y$
2	y_2	2	$2y_2$	2	$4y$	2	$8y$
3	y_3	4	$4y_3$	3	$12y$	3	$36y$
4	y_4	2	$2y_4$	4	$8y$	4	$32y$
5	y_5	4	$4y_5$	5	$20y$	5	$100y$
6	y_6	1	$1y_6$	6	$6y$	6	$35y$
			ΣF (area)		ΣF (First moment)		ΣF (Second moment)

$$\therefore \quad \text{Area} = \frac{h}{3} \Sigma F \text{ (area)}$$

$$\text{First moment (about Station 0)} = \frac{h}{3} h \Sigma F \text{ (first moment)}$$

$$\text{Second moment (about Station 0)} = \frac{h}{3} h^2 \Sigma F \text{ (second moment)}$$

In general, we need to use the area and first moment of the area to find the position of the centroid of the area (or of the volume when volume is being considered). Thus, to find the position of the centroid of the area bout axis YY, we use:

$$\bar{X} = \frac{\text{First moment of area}}{\text{area}}$$

$$= \frac{\dfrac{h}{3} h \Sigma F \text{ (First moment)}}{\dfrac{h}{3} \Sigma F \text{ (area)}} \quad \text{and cancelling } \frac{h}{3} \text{ gives}$$

$$= \frac{h \Sigma F \text{ (First moment)}}{\Sigma F \text{ (Area)}}$$

For work on longitudinal stability, we require the second moment about the transverse axis which passes through the centroid of the area. By making use of the theorem of parallel axes, we can say that:

$$I_{NA} = I_{YY} - Ax^2$$

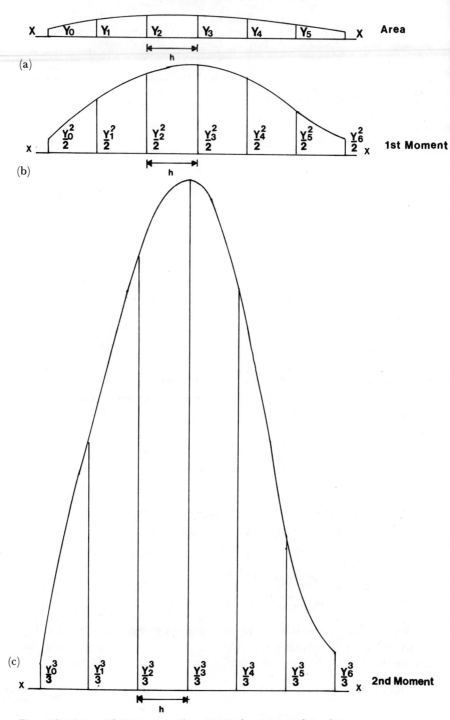

Figure 1.20 (a) Area. (b) First moment of Area (a). (c) Second moment of Area (a)

To find moments about a longitudinal axis

The first and second moments of the ship's waterplane area about its centreline are important values in the calculation of its stability data. These values are determined at the design stage of the vessel and are not normally calculated by the mariner.

However, an understanding of the means by which these figures are obtained is of value and are explained below.

When finding the centroid of the area relative to axis XX, note that the first moment of the strip area $y\,dl$ is $(y/2)y\,dl = (y^2/2)\,dl$ (*Figure 1.21*)

Hence, the first moment of the area $= \displaystyle\int_0^L \frac{y^2}{2}\,dl$

In finding the second moments of the area about an axis, note that the

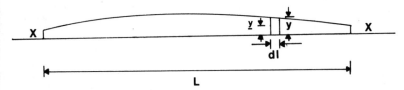

Figure 1.21 Moment about XX

second moment of the rectangular strip $y\,dl$ about axis XX is

$$\frac{y^3\,dl}{3}$$

Hence, the second moment of the area $= \displaystyle\int_0^L \frac{y^3}{3}\,dl$

The effect of translating these quantities into areas is shown in *Figure 1.20*, where (*b*) is first moments and (*c*) is second moments.

To find the first moment of the waterplane area about its centreline (Figure 1.20(b))

Table 1.5

Station	Ord	Ord²	Simpson's multiplier	F (First moment)
0	y_0	y_0^2	1	y_0^2
1	y_1	y_1^2	4	$4y_1^2$
2	y_2	y_2^2	2	$2y_2^2$
3	y_3	y_3^2	4	$4y_3^2$
4	y_4	y_4^2	2	$2y_4^2$
5	y_5	y_5^2	4	$4y_5^2$
6	y_6	y_6^2	1	y_6^2

ΣF (First moment)

and first moment $= \dfrac{h}{3}\dfrac{1}{2}\Sigma\, F$ (First moment)

To find the second moment of the waterplane area about its centreline

Table 1.6

Station	Ord	Ord^3	Simpson's multiplier	F (Second moment)
0	y_0	y_0^3	1	y_0^3
1	y_1	y_1^3	4	$4y_1^3$
2	y_2	y_2^3	2	$2y_2^3$
3	y_3	y_3^3	4	$4y_3^3$
4	y_4	y_4^3	2	$2y_4^3$
5	y_5	y_5^3	4	$4y_5^3$
6	y_6	y_6^3	1	y_6^3

$\Sigma\, F$ (second moment)

and second moment $= \dfrac{h}{3}\dfrac{1}{3}\Sigma\, F$ (second moment)

Note that, for the *whole* waterplane (i.e. both sides)

$$\text{second moment} = 2\,\frac{h}{3}\frac{1}{3}\,\Sigma\, F \text{ (second moment)}$$

DETERMINING VALUES OF FIRST AND SECOND MOMENTS ABOUT AXES FOR SHIP'S WATERPLANES

The procedure for determining the values of first and second moments about transverse and longitudinal axes for ships' waterplanes can best be understood by the examination of a worked example in conjunction with the explanation given above.

The following example serves as an illustration of these methods.

Example 1.4

A vessel of length 200 m has the half ordinates of waterplane values as shown, commencing at the after perpendicular (AP):

Station	0	1	2	3	4	5	6	7	8	9	10
½ Ord	0	10.0	13.0	14.0	14.2	14.2	14.1	14.0	11.5	6.2	0.2

Find:

(a) The area of the waterplane.
(b) The position of the centroid of the waterplane relative to the after perpendicular.
(c) The second moment of the area about a transverse axis through the centroid.
(d) The second moment of the area about a longitudinal axis through the centreline.

Station	½ Ord	Simpson's multiplier	F (area)	Lever (First moment)	F (First moment)	Lever (Second moment)	F (Second moment)
0	0	1	0	0	0	0	0
1	10.0	4	40.0	1	40.0	1	40.0
2	13.0	2	26.0	2	52.0	2	104.0
3	14.0	4	56.0	3	168.0	3	504.0
4	14.2	2	28.4	4	113.6	4	454.4
5	14.2	4	56.8	5	284.0	5	1420.0
6	14.1	2	28.2	6	169.2	6	1015.2
7	14.0	4	56.0	7	392.0	7	2744.0
8	11.5	2	23.0	8	184.0	8	1472.0
9	6.2	4	24.8	9	223.2	9	2008.8
10	0.2	1	0.2	10	2.0	10	20.0
			339.4		1628.0		9772.4

$\therefore$ (a) Area $= 2 \times \dfrac{1}{3} \times h \quad F \text{ (area)}$

$$= 2 \times \frac{1}{3} \times 20 \times 339.4 = 4525.3 \text{ m}^2$$

(b) Centroid about AP $= \dfrac{h \Sigma F \text{ (first moment)}}{\Sigma F \text{ (area)}}$

$$= \frac{20 \times 1628.0}{339.4}$$

$$= 95.93 \text{ m forward of AP}$$

(c) $I_{AP} = 2 \times \dfrac{1}{3} \times h^3 \Sigma F$ (second moment)

$$= \frac{2}{3} \times 20^3 \times 9782.4$$

$$= 52\,172\,800 \text{ m}^4$$

This value must now be modified by obtaining the second moment about the centroid of the waterplane using the parallel axes theorem, as follows:

$$I_{NA} = I_{AP} - A\bar{x}^2$$

$$= 52\,172\,800 - (4525.3 \times 95.92^2)$$

$$= 10\,528\,433 \text{ m}^4$$

(d) To determine the second moment of the waterplane area about the centreline, cube the ordinate values as follows:

Station	½ Ord	½ Ord³	Simpson's multiplier	F (Second moment)
0	0	0	1	0
1	10.0	1000.0	4	4000.0
2	13.0	2197.0	2	4394.0
3	14.0	2744.0	4	10976.0
4	14.2	2863.0	2	5726.0
5	14.2	2863.0	4	11452.0
6	14.1	2803.0	2	5606.0
7	14.0	2744.0	4	10976.0
8	11.5	1520.0	2	3040.0
9	6.2	238.0	4	952.0
10	0.2	0.0	1	0
				Σ 57122.0

$$I_{CL} = 2 \times \frac{1}{3} \times \frac{1}{3} h \Sigma F \text{ (second moment)}$$

$$= 2 \times \frac{1}{3} \times \frac{1}{3} \times 20 \times 57122.0$$

$$= 253\,875.6 \text{ m}^4$$

QUESTIONS ON SIMPSON'S RULES

1. A vessel of length 240 m has the $\frac{1}{2}$ ordinates of waterplane area shown, spaced at equal intervals; commencing from the AP

Station	0	1	2	3	4	5	6	7	8
$\frac{1}{2}$ Ord (m)	0	12.0	14.5	15.0	15.0	14.0	12.0	8.0	3.5

The vessel has a draft of 20 m and the block coefficient at this draught is 0.79 m. Calculate the TPC in salt water and the Fresh Water Allowance at this draft.

2. A vessel has the following dimensions: Length 120.0 m; block coefficient, 0.78; KB 3.2 m.
    $\frac{1}{2}$ ordinates of the waterplane at the 6.0 m draft are:

Station	0	1	2	3	4	5	6
$\frac{1}{2}$ Ord (m)	0	14.0	15.0	15.0	15.0	9.5	1.0

If the vessel is floating on an even keel at a draft of 6.0 m, calculate the vessel's KM.

3. An undivided double bottom tank with vertical sides of depth 1.5 m has $\frac{1}{2}$ ordinates commencing at the after bulkhead as follows:

Station	0	1	2	3	4	5	6
$\frac{1}{2}$ Ord (m)	6.0	6.0	5.9	5.7	5.4	4.9	4.3

The after bulkhead is 65 m forward of the after perpendicular. Find the volume of the tank, the distance of the centroid of the tank from the AP and the free surface moment of the tank when it contains fresh water.

SIMPSON'S RULES (ANSWERS)

1. Σ Function of area $=$ 282.5
    Area $= 5650 \text{ m}^2$
    TPC $=$ 57.9 tonnes/cm
    FWA $=$ 491.2 mm

2. Underwater volume $= 16850 \text{ m}^3$
    Σ Function I_q $= 41405$ $I_q = 184.022 \text{ m}^4$
    BM $=$ 10.95 m
    KM $=$ 14.14 m

3. Σ Function for area   =   99.3
    Σ Function for First mom   =   283.0
    Σ Function for Second mom = 2719.7
    Volume   =   397.2 m^3
    Free surface moment   = 2417.5 tonnes metres

2 Flotation

This chapter deals with the principle of floatation and with its application to ship operations.

OBJECTIVES

1. To enable the pressure and thrust in a liquid to be determined.
2. To enable the effect on suspended and floating bodies of static thrust to be determined (Archimedes' Principle and the Principle of Floatation).
3. To define displacement, deadweight and TPC.
4. To define the relationships between displacement, draft and density.
5. To define the form of ships' load lines.
6. To enable FWA and DWA to be determined.
7. To enable the amount of cargo to be loaded or discharged when limited by load line considerations to be determined.

PRESSURE

Consider an area A located at a depth h below the surface of a liquid of density ρ (see *Figure 2.1*). The force exerted by the liquid over this area A may be determined as follows:

$$F = \text{Volume} \times \rho \times g \quad (F \text{ is the force acting on the area;}$$
$$= Ah\rho g \quad \quad \quad \quad \quad (g \text{ is the acceleration due to gravity)}$$

The pressure acting on the area is expressed as the force per unit area, i.e.

$$\text{Pressure} = \frac{\text{force}}{\text{area}}$$

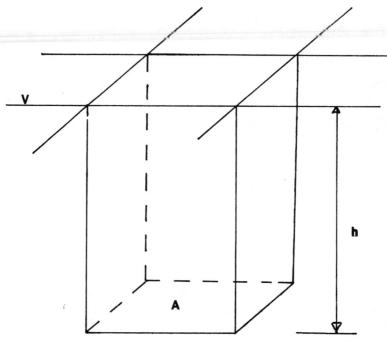

Figure 2.1 Pressure in a liquid

$$\therefore \quad P = \frac{Ah\rho g}{A} = h\rho g$$

i.e.   Pressure = depth × density × g

ARCHIMEDES' PRINCIPLE

This states that a body wholly or partially immersed in a liquid is subject
to an upthrust equal to the weight of liquid displaced by the body.

Buoyancy and floatation

Consider a body with straight sides, of depth d and cross-sectional area A
with the upper surface parallel to the surface of the liquid in which it is
immersed and a distance h below that surface (see *Figure 2.2*). The density

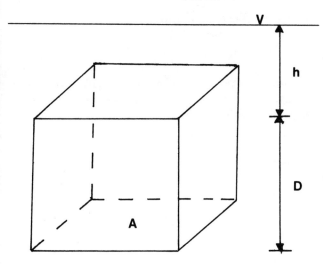

Figure 2.2 Archimedes' principle

of the liquid is ρ. Then:

Thrust on the upper surface $= Ah\rho g$

Thrust on the lower surface $= A(h+d)\rho g$

$\therefore$ The upthrust on the body $\quad = A(h+d)\rho g - Ah\rho g$
$$= Ad\rho g$$

Also, the weight of the liquid which is displaced by the body immersed in it

$$= Ad\rho g$$

Hence, upthrust on the body $=$ weight of liquid displaced.

PRINCIPLE OF FLOATATION

It is evident that, when objects are placed in a liquid, some of them will sink and others will float. We must determine the condition which will result in an object or body floating when placed in a liquid. From the previous section, we established that the weight of the liquid displaced by the immersed body was equal to $Ad\rho g$. If the liquid has a density of ρ_1, this may be expressed as:

Weight of liquid displaced $= Ad\rho_1 g$

Also, the weight of the body, if it has a density ρ_B, may be expressed as:

Weight of body $= Ad\rho_B g$

If the weight of the immersed body is equal to the weight of the liquid displaced by the body, then the body will not move up or down. Then occurs when $\rho_1 = \rho_B$.

i.e. $Ad\rho_1 g = Ad\rho_B g$

When $\rho_1 < \rho_B$, then

$Ad\rho_1 g < Ad\rho_B g$

and the body will sink.
   When $\rho_1 > \rho_B$, then

$Ad\rho_1 g > Ad\rho_B g$

and the body will be lighter than the weight of liquid which it displaces. As the upthrust acting on the fully immersed body is greater than the weight of the body, the body will rise to the surface of the liquid and will float with part of its volume out of the liquid, i.e. it will rise until (*Figure 2.3*):

Weight of the body $=$ weight of the liquid displaced
by the body

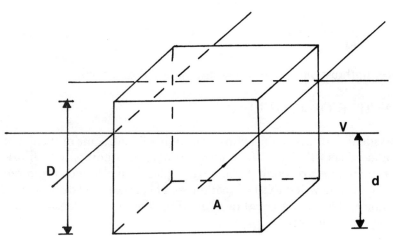

Figure 2.3 Principle of flotation

This means that, if the depth of the body is D and the draft at which it floats with its area parallel to the liquid surface is d, the body will rise until:

$$Ad\rho_1 g = Ad\rho_B g$$

(*Note.* The quantities considered above are strictly forces and are measured in Newtons. However, in most of the operational work concerning the buoyancy and stability of floating bodies, mass units are most convenient. For the majority of the work of this book, therefore, tonnes will be used, where 1 tonne is 1000 kg.)

DISPLACEMENT

Since a floating body displaces its own weight of water, the most convenient method of finding the mass of an irregular shaped body floating in a liquid is to take the product of the underwater volume and the density of the liquid in which it floats. The mass of the floating body is generally termed its displacement, i.e.

Displacement = underwater volume × ρ_1

(where ρ_1 is again the density of the liquid in which the body is floating).
   For a box shape of length L and breadth B, floating at draft d in a liquid of density ρ_1, then:

Displacement = $L \times B \times d \times \rho_1$

For shapes of ships, the underwater volume can be found using Simpson's Rules (see chapter 1). When this volume has been found, it can be compared to the volume of a rectangular shape having the same length, breadth and draft as the vessel. The ratio of the underwater volume of the ship to the volume of the block having the same length, breadth and draft is termed the block coefficient of the vessel, C_b, i.e. (*Figure 2.4*)

$$C_b = \frac{\text{underwater volume}}{L \times B \times d}$$

C_b is a measure of the 'fineness' of the vessel's underwater form and is often used in the application of regulations of vessels, such as those governing its freeboard or strength.

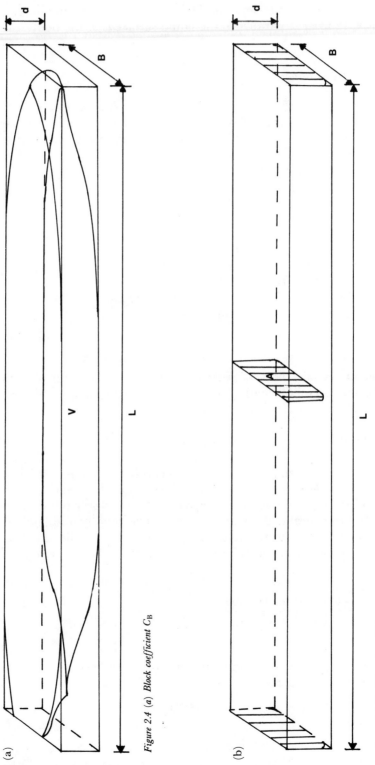

Figure 2.4 (a) Block coefficient C_B

Figure 2.4 (b) Prismatic coefficient

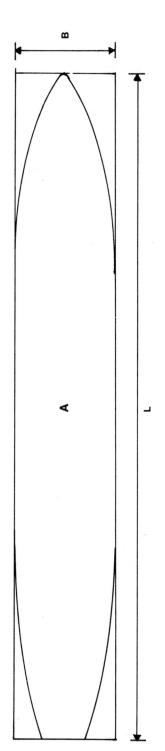

Figure 2.4 (c) Coefficient of fineness of waterplan C_W

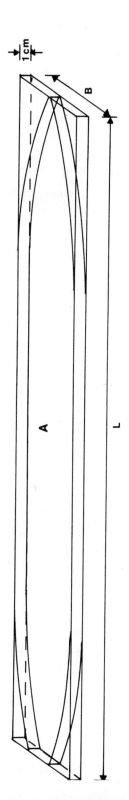

Figure 2.5 Tonnes per centimetre immersion

It should be noted that, when the C_b of a ship is known, the underwater volume of the ship can be found using:

$$\text{Volume} = L \times B \times d \times C_b$$

and hence the ship's displacement can be obtained for this draft and C_b value.

Changes in displacement

The effects of changes in displacement, draft and density of the liquid in which the vessel is floating can be examined. For box shapes, we know that:

$$\text{Displacement}_1 = L \times B \times d_1 \times \rho_1$$

With constant density, but at a new draft (d_2):

$$\text{Displacement}_2 = L \times B \times d_2 \times \rho_1$$

$$\therefore \quad \frac{\text{Displacement}_1}{\text{Displacement}_2} = \frac{d_1}{d_2}$$

i.e. the draft is directly proportional to displacement.

When no changes are made to the mass of the ship, its displacement remains constant. However, draft changes may result at constant displacement due to changes in the density of the liquid in which the vessel floats; i.e.

$$\text{Displacement} = L \times B \times d_1 \times \rho_1$$

$$\text{Displacement} = L \times B \times d_2 \times \rho_2$$

$$\therefore \quad L \times B \times d_1 \times \rho_1 = L \times B \times d_2 \times \rho_2$$

i.e. $\quad \dfrac{d_1}{d_2} = \dfrac{\rho_2}{\rho_1}$

which is true for box shapes only.

As the range of densities in which a vessel (ship shape) will float is small; by convention the density of fresh water is 1.000 tonnes/m^3 and the density of salt water is taken to be 1.025 tonnes/m^3; then as the changes of block coefficient over the range of drafts due to the changes in density will be negligible. For ship shapes, therefore, we can say that:

$$\frac{d_1}{d_2} = \frac{\rho_2}{\rho_1}$$

TONNE PER CENTIMETRE IMMERSION

Changes in draft cause changes in displacement and in calculating displacement changes due to draft changes it is useful to have a value for the number of tonnes which will cause the draft of a ship to change by 1 cm.

This figure is termed the Tonnes per Centimetre (TPC) and may be applied to the calculation of draft changes due to added weights (immersion) or discharged weights (emergence) (*Figure 2.5*).

Consider a vessel of waterplane area A at a certain draft, floating in water of density ρ_1. If a weight w tonnes is loaded onto the vessel, causing it to sink by 1 cm, then the mass of the added weight must cause the displacement of an equal mass of the liquid, resulting in the vessel sinking by the 1 cm, i.e. the mass of the displaced liquid may be found using:

$$\text{Mass} = A \times \frac{1}{100} \times \rho_1 \frac{\text{tonnes}}{\text{cm}} \quad \left(1 \text{ cm} = \frac{1}{100} \text{ m}\right)$$

Since w is the weight which will sink the vessel by 1 cm:

$$w = \text{TPC}$$

i.e. $\text{TPC} = \dfrac{A\rho_1}{100} \dfrac{\text{tonnes}}{\text{cm}}$

and if the vessel is floating in salt water of density 1.025 tonnes/m^3, then:

$$\text{TPC} = \frac{A \, 1.025}{100} \frac{\text{tonnes}}{\text{cm}}$$

Example 2.1

A baulk of timber 4 m long, 1.5 m wide and 1 m deep has a relative density (RD) of 0.75. A steel cube of side 0.5 m and RD 8.0 is suspended beneath the timber and the two float in salt water of RD 1.025. Calculate the draft of the timber baulk.

Mass of floating body = mass of liquid displaced

Mass of timber + mass of steel

$$- \text{mass of liquid displaced by timber}$$
$$+ \text{mass of liquid displaced by steel}$$

i.e. $1.5 \times 1 \times 4 \times 0.75 + 0.5^3 \times 8.0 = 1.5 \times 4 \times d + 0.5^3 \times 1.025$

∴ $\quad 4.5 \text{ tonne} + 1.0 \text{ tonne} = 6.150d \text{ tonne} + 0.128 \text{ tonne}$

$$5.5 - 0.128 = 6.150d$$

∴
$$d = \frac{5.372}{6.150}$$

$$= 0.872 \text{ m}$$

Example 2.2

A buoy is to be used to float a mooring chain. The buoy is cylindrical, of length 4 m, radius 0.75 m and is to float half submerged.

The mooring chain is made of steel of RD 7.95 and has a mass of 0.15 tonnes/m. The chain will be suspended in 10 m of salt water.

Calculate:

(a) The mass of the buoy;
(b) The thickness of the plate which should be used to construct the buoy if the buoy is to be made of metal of RD 7.5.

(a) Mass of buoy + mass of chain = mass displaced by buoy
$$\qquad\qquad\qquad\qquad\qquad\qquad + \text{mass displaced by chain}$$

i.e. $\qquad\qquad M_b + 10 \times 0.15t = \frac{1}{2} \times 0.75^2 \times 4 \times 1.025$

$$+ \frac{10 \times 0.2}{7.95} \times 1.025$$

$$M_b + 1.5t = 3.623t + 0.258t$$

∴ $\qquad\qquad\qquad\qquad M_b = 2.380 \text{ tonnes}$

(b) Let the thickness of the buoy metal be t metres:

$$M_b = v \times \rho$$

where v is the volume of the metal and ρ is the density of the metal

∴ $\quad 2.380 = (2\pi r t L + 2\pi r^2 t)7.95$
$$= 7.95 \times 2\pi r t(L + r)$$
$$= 7.95 \ 2\pi \ 0.75 \ 4.75t$$
$$= 177.952t$$

$$\therefore \qquad t = \frac{2.380}{177.052} = 0.0134 \text{ m}$$

i.e.   thickness of metal = 13.4 mm

DEFINITIONS

When examining the displacements which a ship may have at various drafts, two displacements in particular are of particular significance, namely:

Light Displacement, which is the weight of liquid displaced by a vessel when floating with no cargo, fuel, stores or any other weights not forming part of the hull or machinery or fixed equipment of the vessel, but including water in the boilers and condensers to working level.
Load Displacement, which is the weight of liquid displaced by a vessel when floating on an even keel at her summer load draft in salt water.

These two displacements are the lower and upper limits of displacement between which the vessel would normally be expected to work. The difference between the two is the useful carrying capacity of the vessel, i.e. the amount of cargo, fuel, stores, ballast etc. which, when added to her light displacement, will bring the vessel to her load displacement. This difference is termed the ship's 'deadweight'.

LOAD LINES

Load lines are marked on a vessel's side at mid-length to define the maximum drafts to which the vessel may load in all sea areas and also in rivers and harbours where the density of the water in which the vessel floats is not equal to the density of salt water.

This book is not concerned with the details of the method by which the freeboard of a vessel is determined. It is sufficient to say here that the vessel must first be deemed to be fit to be assigned a freeboard and then the summer freeboard in salt water is determined. The summer freeboard depends upon:

The length of the vessel;
The proportions of the vessel's principal dimensions;
The block coefficient of the vessel;
The amount of superstructure on the vessel;
The degree of watertight as distinct from weathertight integrity of the vessel;
The height of the bow.

After the summer freeboard has been determined, all other freeboards are based upon this value.

It should be noted that a small proportion of ships in the world fleet are assigned lumber (or timber) load lines. These permit the vessel to have a freeboard which is less than her 'normal' summer freeboard when she is carrying a certain quantity of properly stowed timber deck cargo.

The difference between the summer freeboard and the lumber summer freeboard is essentially an improved allowance for the ship's super-structures based on the assumption that the timber effectively adds to the ship's superstructure volume. The reasons for this additional allowance are discussed in Chapter 6.

Having established the summer freeboard, the relationships between this and the other load line marks are laid down in load line regulations, as given below (*Figure 2.6*).

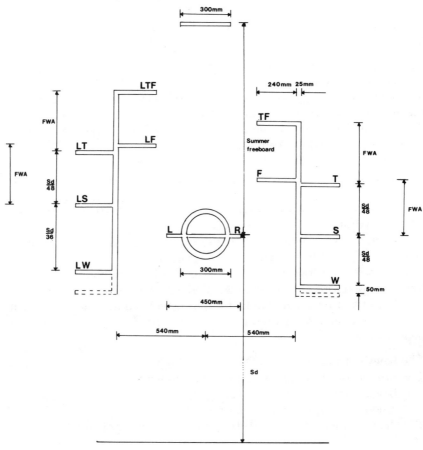

Figure 2.6 Load lines

Summer freeboard
The distance in millimetres from the top of the deck line to the top of the line passing through the centre of the load line disc (and to the top of the line forward of the disc).

Tropical load line
A line 1/48th of the summer draft of the vessel above the summer load line.

Winter load line
A line 1/48th of the summer draft below the summer load line.

Winter North Atlantic load line
(This only applies to vessels less than 100 m long.) A line 50 mm below the winter load line. Note that this line is not marked on larger vessels.

Summer fresh water load line
A line drawn a distance equal to the fresh water allowance above the summer load line.

Tropical fresh water load line
A line drawn a distance equal to the fresh water allowance above the tropical load line.

For those vessels marked with lumber load lines, the following relationships between the various load lines have been established.

Lumber summer load line
A line drawn a distance equal to the lumber summer freeboard below the deck line.

Lumber tropical load line
A line 1/48th of the lumber summer draft above the lumber summer load line.

Lumber winter load line
A line 1/36th of the lumber summer draft below the lumber summer load line.

Lumber fresh water load line
A line drawn a distance equal to the lumber fresh water allowance above
the summer load line.

Lumber tropical fresh water load line
A line drawn a distance equal to the lumber fresh water allowance above
the lumber tropical load line.

Lumber Winter North Atlantic load line
In vessels of less than 100 m in length, this load line is drawn at the same
level as the normal winter north Atlantic load line, i.e. no allowance is
given to vessels of less than 100 m in length for the deck cargo when
operating in the north Atlantic in winter.

Note that all dimensions are measured to the *top* of the relevant load
line.

The letters marked on either side of the load line disc indicate the
identity of the Assigning Authority which is responsible for implementing
the load line rules. These letters will identify the Classification Society or
other authority, as in the illustration of a typical load line marking in the
diagram below, where Lloyd's Register is the Assigning Authority.

Fresh Water Allowance (FWA)

The Fresh Water Allowance (*Figure 2.7*) is the amount by which the vessel
will change her draft when moving from salt water of density

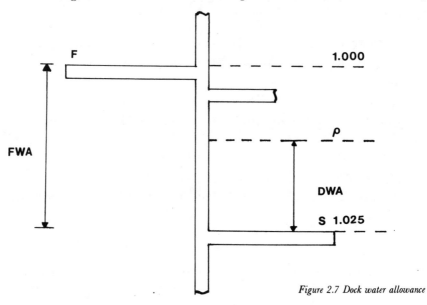

Figure 2.7 Dock water allowance

1.025 tonnes/m^3 to fresh water of density 1.000 tonnes/m^3, and vice versa.

If the salt water density is denoted ρ_s and the fresh water density ρ_f, then a vessel floating at draft d_s in salt water moving into fresh water will float at a new draft $d_s + \text{FWA}$. Now we have that:

$$\frac{d_s + \text{FWA}}{d_s} = \frac{\rho_s}{\rho_f}$$

i.e. $d_s + \text{FWA} = d_s \dfrac{\rho_s}{\rho_f}$

$\therefore$ $\text{FWA} = d_s \dfrac{\rho_s}{\rho_f} - d_s$

$$\text{FWA} = d_s \left(\frac{\rho_s}{\rho_f} - 1\right) = d_s \left(\frac{1.025}{1.000} - 1\right)$$

$\therefore$ $\text{FWA} = 0.025 d_s$ m

or $\text{FWA} = 25 d_s$ mm

A fresh water allowance can be calculated for a ship at any draft at which she is floating. However, it is normally only of interest when the ship is at, or close to, her summer load line and it is desired to determine how much deeper than the summer draft in salt water she may go when she is operating in fresh water.

The fresh water allowance stated in the vessel's load line calculation is only applicable when the vessel is operating at or near her summer load draft, i.e. when d_s is the summer load draft in salt water. In which case:

$$\text{FWA} = 25 d_s \frac{A \times 1.025 \times 10^{-2}}{A \times 1.025 \times 10^{-2}}$$

where A is the waterplane area at the summer draft. As $A \times 1.025 \times 10^{-2} = \text{TPC}$ and $d_s \times A \times 1.025$ approximates to the summer displacement:

then $\text{FWA} = \dfrac{25 \times \text{displacement}}{100 \times \text{TPC}}$ mm

i.e. $\text{FWA} = \dfrac{\text{displacement}}{4 \text{ TPC}}$

Dock Water Allowance (DWA)

The Dock Water Allowance (*Figure 2.8*) is the amount by which the vessel will change her draft when moving from salt water of density 1.025 tonnes/m³ to water of a different density in between the density of salt water and that of fresh water, or vice versa. With the salt water density of 1.025 tonnes/m³ and the dock water density denoted ρ_D, the dock water allowance is calculated as follows:

$$\frac{DWA}{FWA} = \frac{1.025 - \rho_D}{1.025 - 1.000}$$

$$DWA = FWA \frac{(1.025 - \rho_D)}{1.025 - 1.000}$$

i.e. $DWA = FWA \dfrac{(1.025 - \rho_D)}{0.025}$

For ease of manipulation of the numbers, the top and bottom lines are multiplied by 1000. This means that the density of the dock water will lie

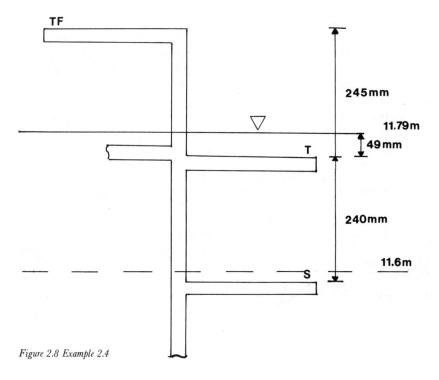

Figure 2.8 Example 2.4

between the values of 1000 and 1025, giving the formula:

$$DWA = FWA \frac{1025 - \rho_D}{25}$$

The dock water allowance, like the fresh water allowance, is normally only of importance where the vessel is operating at or near her summer load draft. In cases where the vessel is at some lesser d t it is possible to calculate the dock water allowance, making us the approximation:

$$\frac{d_1}{d_2} = \frac{\rho_1}{\rho_2}$$

Example 2.3

A vessel of summer displacement 30 000 tonnes has a summer load draft of 11.5 m, TPC 30.6 tonnes/cm. If the vessel is floating at a draft of 11.6 m in water of density 1.020 tonnes/m^3 in a tropical zone, how much more cargo may she load to bring her to her summer load line when floating in salt water?

Solution. See *Figure 2.8*
Allowance between summer and tropical load lines

$$= \frac{\text{summer draft}}{48} \times 1000 \ (\times 1000 \text{ to convert to mm})$$

$$= \frac{11.5}{48} \times 1000 = 240 \text{ mm}$$

As $\quad FWA = \dfrac{W}{4 \ TPC} \qquad$ then $\qquad FWA = \dfrac{30\,000}{4 \times 30.6} = 245 \text{ mm}$

As $\quad DWA = FWA \dfrac{(1025 - \rho_D)}{25}$

$\therefore \quad DWA = 245 \dfrac{(1025 - 1020)}{25}$

$$= 245 \times \frac{5}{25} = 49 \text{ mm}$$

Knowing the summer to tropical allowance and the DWA, the maximum permitted draft may now be calculated as follows:

Summer draft	= 11.500 m	
S-T allowance	= 0.240 m	
DWA	= 0.049 m	
Max. draft	= 11.789 m	(in dock water in the
Present draft	= 11.600 m	tropical zone)
Permitted sinkage	= 0.189 m	
i.e. Permitted sinkage	= 19 cm	

Knowing the sinkage, the amount of cargo to load can now be determined by multiplying by the TPC of the vessel. However, unless otherwise stated, the TPC value quoted is given for salt water. As the vessel is floating in dock water of a different density to that of salt water, the TPC value should be corrected, as follows:

$$TPC_D = TPC_S \frac{\rho_D}{\rho_S}$$

where D and S denote dock and salt values respectively.

$$\therefore \quad TPC_D = 30.6 \frac{1020}{1025} = 30.45 \text{ tonnes/cm}$$

$\therefore$ Cargo to be loaded = sinkage × corrected TPC

$$= 19 \times 30.45 = 578.6 \text{ tonnes}$$

The permitted sinkage of 18.9 cm has been rounded to 19 cm, giving a small difference in the final answer. It will be appreciated that the draft of the vessel originally stated (11.6 m) could not be read to an accuracy of 1 mm and therefore rounding of the permitted sinkage to the nearest cm is reasonable.

Example 2.4

A vessel about to complete loading in a summer zone is expected to enter a winter zone after steaming from the loading port for 10 days. On passage, fuel consumption is expected to be 30 tonnes per day and water consumption 15 tonnes per day.

The ship is at present floating in water of density 1.013 tonnes/m³ at a draft of 9.0 m.

Summer load draft	9.475 m
FWA	203 mm
TPC	30 tonnes/cm

Find

(a) The maximum permissable draft on completion of loading.
(b) The amount of cargo which the vessel can load.
(Note. A vessel passing from a summer zone to a winter zone, or from a tropical zone to a summer zone, must arrive at the zone boundary at the draft appropriate to that zone.)

Solution

(a) Allowance between summer and winter load lines

$$= \frac{\text{summer draft}}{48} \times 1000$$

$$= \frac{9.475}{48} \times 1000 = 197 \text{ mm}$$

As $\quad \text{DWA} = \text{FWA} \dfrac{(1025 - \rho_D)}{25}$

$$\text{DWA} = 203 \frac{(1025 - 1013)}{25} = 97 \text{ mm}$$

To calculate the change of draft while on passage to the boundary of the winter zone, the total consumption of fuel and water must be calculated and divided by TPC, as follows:

$$\text{Change of draft on passage} = \frac{30 \times 10 + 15 \times 10}{30} = 15 \text{ cm}$$

Knowing the DWA and the allowance to make for fuel and water consumed on passage to the zone boundary, the maximum permitted draft may now be calculated, as follows:

Summer draft	= 9.475 m
S-W allowance	= 0.197 m

Winter draft	= 9.278 m
DWA draft	= 0.097 m
'In dock' draft	= 9.375 m
Passage allowance	= 0.150 m

Max. draft	= 9.525 m
Present draft	= 9.000 m

Permitted sinkage	= 0.525 m

The calculation so far has established that the vessel may load to a draft of 9.525 m in dock in order to be at the appropriate winter draft on arrival at the winter zone boundary. There is another limitation which must be considered, i.e. will the vessel be deeper than the summer draft plus the DWA when at this draft in the dock? If so, then the figure of 9.525 m will have to be reduced accordingly and the permitted sinkage will similarly decrease.

Summer draft	= 9.475 m
DWA	= 0.097 m

Max. draft	9.572 m

As this figure is greater than 9.525 m, then 9.525 is clearly the limiting draft to which the vessel can be loaded in the dock. The amount of cargo to load must now be determined using the permitted sinkage of 0.525 m or 52.5 cm, again correcting the TPC for the density of the dock water, as follows:

(b) $\quad TPC_D = TPC_S \dfrac{\rho_D}{\rho_S}$

$$= 30 \times \frac{1013}{1025} = 29.65 \text{ tonnes/cm}$$

$\therefore \quad$ Cargo to load $= \text{sinkage} \times TPC_D$
$$= 52.5 \times 29.65$$
$$= 1556.6 \text{ tonnes at maximum draft}$$
$$\text{of } 9.525 \text{ m}$$

Example 2.5

A vessel has a summer freeboard of 4.2 m, which corresponds to a draft of
8.9 m. Her FWA is 178 mm and TPC 22.5 tonnes/cm.

   The vessel is floating in river water of RD 1.010 in a tropical zone, with
present freeboards (measured at midlength) of 4.4 m to starboard and
4.3 m to port. It is estimated that the vessel will use 70 tonnes of fuel and 5
tonnes of fresh water on her passage down river to the sea.

   What is the maximum amount of cargo which she may load in the
river?

Solution
Allowance between summer and tropical load lines

$$= \frac{\text{summer draft}}{48} \times 1000$$

$$= \frac{8.9}{48} \times 1000 = 185 \text{ mm}$$

As $DWA = FWA \dfrac{(1025 - \rho_D)}{25}$

$$= 178 \frac{(1025 - 1010)}{25} = 107 \text{ mm}$$

The present draft of the vessel must now be determined by adding
together the summer draft and summer freeboard to obtain the freeboard
depth of the vessel, then subtracting the present freeboard of the vessel
from this depth, as follows:

$$\text{Present draft} = \text{freeboard depth} - \text{mean freeboard}$$

$$= (4.2 + 8.9) - \frac{(4.4 + 4.3)}{2}$$

$$= 8.750 \text{ mm}$$

Knowing the summer to tropical allowance and the DWA, the maximum
permitted draft may now be calculated, as follows:

Summer draft	$= 8.900$ m
S-T allowance	$= 0.185$ m
DWA	$= 0.107$ m
Max. draft	$= 9.192$ m
Present draft	$= 8.750$ m
Permitted sinkage	$= 0.292$ m

Again, the TPC value stated is assumed to be for salt water and must be corrected to obtain the correct value for river water of a different density, as follows:

$$\text{TPC}_D = \text{TPC}_S \frac{\rho_D}{\rho_S}$$

$$= 22.5 \times \frac{1010}{1025}$$

$$= 22.2 \text{ tonnes/cm}$$

$\therefore$ to be loaded in the river
= permitted sinkage $\times$ TPC + consumption
of fuel and water on passage to the sea
$= (29.2 \times 22.2) + (70 \times 5)$
$= 731$ tonnes

QUESTIONS ON HYDROSTATICS

1. For the purposes of current measurement, buoys are to be used which consist of a long pole and a ballast weight. In order to minimise wind effect, the buoys are to float with only a short length above the water.

   The buoy consists of a pole 2 m long and 20 mm in diameter, of RD 0.7, and a ballast weight of steel of RD 8.0.

   If the buoy is to float with exactly 30 mm of the pole above the water, how much ballast is needed in fresh water (RD 1.000)?

2. A sphere of brass of RD 3.4 and radius 0.1 m rests on the bottom of a tank containing oil of RD 0.8. A wooden sphere of RD 0.5 is attached by a string to the immersed sphere.

   Find the minimum diameter of the wooden sphere to just cause the brass ball to float clear of the bottom of the tank.

3. A box shaped vessel has a length of 100 m, breadth of 10 m and draft of 4 m in water of density 1.010 tonnes/m^3. Find:
   (a) Her present displacement;
   (b) Her new draft if she loads 750 tonnes of cargo in this density of water;
   (c) Her new draft if she then proceeds to sea (density 1.025 tonnes/m^3);
   (d) Her new draft on arrival at a port where the water density is 1.005 tonnes/m^3 (cargo still on board);
   (e) How much cargo has been discharged at that port if after discharging the draft is found to be 3.5 m.

HYDROSTATICS (ANSWERS)

1. 0.205 kg.

2. 0.205 m.

3. (a) 4040 tonnes; (b) 4.753 m; (c) 4.673 m; (d) 4.767 m; (e) 1272.5
    tonnes.

LOAD LINE QUESTIONS

1. A vessel of summer displacement 17 000 tonnes is floating at a draft of
    9.8 m in water of density 1.008 tonnes/m^3 in a winter zone, TPC 20.4
    tonnes/cm, summer load draft 10.4 m.
      How much more cargo may she load in order to complete loading
    at the appropriate load line?

2. A vessel is floating in dock water of density 1.009 tonnes/m^3 at a mean
    draft of 8.6 m with TPC 2.06 tonnes/cm. The maximum draft over a
    bar, where it is anticipated that the vessel will be floating in salt
    water, is 9.5 m. At a draft of 9.5 m in salt water, the vessel has a
    displacement of 12 000 tonnes.
      If the vessel has yet to load 1500 tonnes of cargo, how many more
    tonnes of fresh water can she take on board in the dock in order to
    ensure that the maximum draft crossing the bar is not exceeded?

3. A vessel displacing 16 000 tonnes has a TPC 20 tonnes/cm and is
    floating at a draft of 9.4 m in water of density 1.010 tonnes/m^3 in a
    summer zone. If the vessel's summer load draft is 9.5 m, calculate the
    amount of cargo she may load.

4. A vessel about to complete loading has the following particulars:

    | Summer load draft | 12.75 m |
    | Summer displacement | 79 000 tonnes |
    | TPC | 75 tonnes/cm |

    The vessel is at present floating in water of density 1.008 tonnes/m^3
    with the bottom of her winter marks 3 cm above the waterline to plrt
    and 15 cm above the waterline to starboard.
      If the vessel is in a tropical zone, how much more cargo may she
    load to bring her to the appropriate maximum draft?

LOAD LINE (ANSWERS)

1. FWA, 208 mm   DWA, 141 mm
   Allowance Summer to Winter, 217 mm
   Cargo to load, 1051 tonnes.

2. FWA, 145.6 mm   DWA, 93 mm
   Fresh water to take on board, 513 tonnes.

3. FWA, 200 mm   DWA, 120 mm
   Sinkage, 22 cm   Cargo to load, 433.6 tonnes.

4. FWA, 263 mm   DWA, 179 mm
   Sinkage, 83 cm   Cargo to load, 6092.0 tonnes.

3 Determination of position of centre of gravity

This chapter deals with finding the position of the centre of gravity when the positions of all the components weights which compose a vessel are known.

OBJECTIVES

1. Give the positions of weights relative to the keel, find the position of the centre of gravity relative to the keel (KG).
2. Given the positions of weights relative to the centreline, find the position of the centre of gravity relative to the centre line.
3. Find the effect of loading, discharging or shifting a single weight.
4. Find the distribution of weight to produce a given position of the centre of gravity.

POSITION OF CENTRE OF GRAVITY RELATIVE TO KEEL AND CENTRE LINE

In *Figure 3.1*, W is the resultant of forces w_0, w_1, w_2 acting at distances x_0, x_1, x_2 from axis YY. Then if the resultant W is a distance x from YY and:

$$W = w_0 + w_1 + w_2$$

$$Wx = w_0 x_0 + w_1 x_1 + w_2 x_2$$

$$x = \frac{w_0 x_0 + w_1 x_1 + w_2 x_2}{W}$$

$$x = \frac{\text{moment of forces about YY}}{\text{weight}}$$

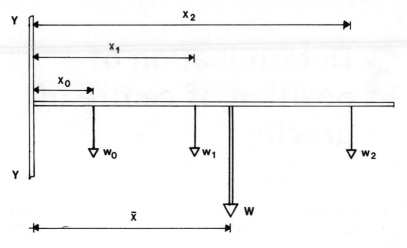

Figure 3.1 Position of centre of gravity

In *Figure 3.2* we consider a ship and take moments about the keel K. In this case

$$W = \text{displacement acting at the centre of gravity } G$$

$$w_0, w_1, w_0 = \text{weights making up the light ship cargo etc. acting at } g_0, g_1, g_2$$

$$W \times KG = w_0 \times Kg_0 + w_1 \times Kg_1 + w_2 Kg_2$$

$$KG = \frac{w_0 \times Kg_0 + w_1 \times Kg_2 + w_2 Kg_2}{W}$$

$$KG = \frac{\text{moments about keel}}{\text{displacement}}$$

Similarly, in *Figure 3.3*, taking moments about the centre line

$$W \times GG + w_0 \times Cg_0 = w_1 \times Cg_1 + w_2 \times Cg_2$$

$$W \times CG = (w_1 \times Cg_1 + w_2 \times Cg_2) - w_0 \times Cg_0$$

$$CG = \frac{(w_1 \times Cg_1 + w_2 \times Cg_2) - w_0 Cg_0}{W}$$

$$CG = \frac{\text{moments of weight about centre line}}{\text{displacement}}$$

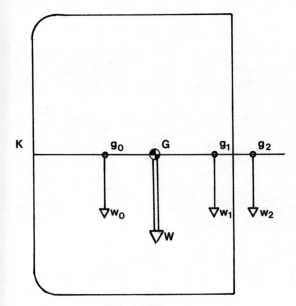

Figure 3.2 KG

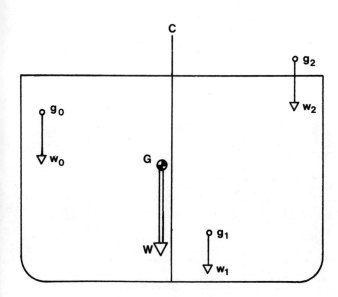

Figure 3.3 G from centreline

SHIFT OF WEIGHT

The vessel in *Figure 3.4* has centre of gravity at G_0 and displacement W. A small weight w is already onboard with its centre of gravity at g_0. The weight w is moved a distance D metre to g_1. The centre of gravity of the vessel moves to G_1.

If the moments of all weights about the keel apart from w is C then

$$W \times KG_0 = C + w \times Kg_0$$

and
$$W \times KG_1 = C + w \times Kg_1$$

$$W \times KG_1 - W \times KG_0 = wKg_1 - w \times Kg_0$$

$$W(KG_1 - KG_0) = w(Kg_1 - Kg_0)$$

$$W \times G_0G_1 = w \times d$$

$$G_0G_1 = \frac{w \times d}{W}$$

ADDING OR REMOVING A SINGLE WEIGHT

If we now consider the situation in *Figure 3.5* where a single weight is loaded at Kg_0 a distance d metres above G_0. Firstly we assume that the weight is loaded at height KG_0 above the keel

$$KG_0(W + w) = C + w \times KG_0$$

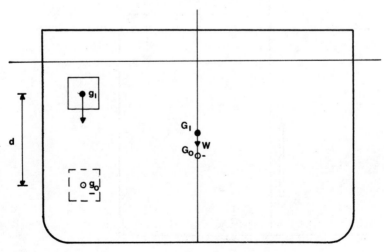

Figure 3.4 Shift of weight

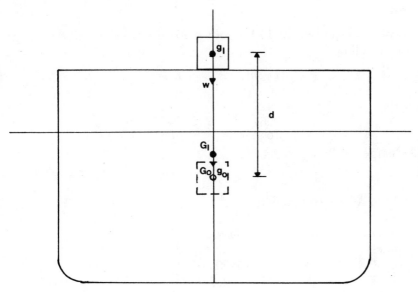

Figure 3.5 Addition of weight

and then moved vertically a distance d metres

$$KG_1(W + w) = C + w \times Kg_1$$

$$KG_1(W + w) - KG_0(W + w) = w \times KG_0 - w \times Kg_1$$

$$(W + w)G_0G_1 = w(KG_0 - Kg_1)$$

$$G_0G_1 = \frac{w \times d}{W + w}$$

These relationships apply equally to transverse and longitudinal shifts of weight. They can also be used to calculate diagonal shifts when the weight shifts vertically and horizontally. The relationship can also be applied to shifts of buoyancy, i.e. when a wedge of buoyancy moves across a vessel, as she is heeled by an external force, when a layer of buoyancy is added as draft increases or when buoyancy is lost as a result of flooding.

These relationships are very useful in deriving formula and describing changes in stability. However, students are warned that they need careful handling and should not be used when more than one operation is involved.

Example 3.1

A vessel of displacement 16 450 tonnes KG 9.3 m loads and discharges cargo as follows:

	Weight (tonnes)	KG
Loads	1427	8.6
	2964	4.6
	1930	12.0
Discharges	2000	11.8
	483	6.4

Find the KG on completion

Weight (tonnes)	KG	Moment
16450	9.3	152985.0
1427	8.6	12272.2
2964	4.6	13634.4
1930	12.0	23160.0
− 2000	11.8	− 23600.0
− 483	6.4	− 3091.2
20288		175.358.4

$$KG = \frac{\text{moment of weight}}{\text{displacement}} = \frac{175358.4}{20288} = 8.643 \text{ m}$$

Example 3.2

A vessel has displacement 6200 tonnes KG, 8.0 m. Distribute 9108 tonnes of cargo between spaces Kg, 0.59 m and 11.45 m, so that the vessel completes loading with a KG of 7.57 m

Load w tonnes at Kg 11.45 m.

Weight (tonnes)	KG	Moment
6200.0	8.00	49600.0
9108.0 − w	0.59	5373.7 − 0.59w
w	11.45	+ 11.45w
15308.0		54973.7 − 10.86w

$$KG = \frac{\text{moment of weight}}{\text{displacement}}$$

$$7.57 = \frac{54973.7 - 10.86w}{15308.0}$$

$$115881.6 = 54973.7 - 10.86w$$

$$w = 5612 \text{ tonnes}$$

Load 5612 tonnes at Kg 11.45 m.
Load 3496 tonnes at Kg 0.59 m.

Example 3.3

A vessel displacement 12500 tonnes has KG 9.6 m. On completion of loading she is required to have KG of 9.5 m.

1000 tonnes is loaded at Kg 5.5 m.
 850 tonnes is loaded at Kg 13.6 m.

Find the Kg at which to load a further 1600 tonnes of cargo to produce the required final KG.
    Let x be the required Kg of 1600 tonnes.

Weight (tonnes)	KG	Moment
12500	9.6	120000
1000	5.5	5500
850	13.6	11560
1600	x	$1600x$
15950		$137060 + 1600x$

$$KG = \frac{\text{moment of weight}}{\text{displacement}}$$

$$9.5 = \frac{137\,060 + 1600x}{15\,950}$$

$$151\,525 = 137\,060 + 1600x$$

$$1600x = 14\,465$$

$$x = 9.041 \text{ m}$$

Load 1600 tonnes at Kg 9.041 m.
(Note. A more practical view is to regard 9.041 m as the mean Kg of the 1600 tonnes after it has been distributed between available spaces.)

QUESTIONS ON KG

1. A vessel has displacement 16 000 tonnes and KG, 8.5 m. She loads the following cargo:

Weight (tonnes)	Kg (m)
1360	4.7
2957	10.5
1638	5.9
500	14.8

Find the KG of the vessel on completion.

2. A vessel has displacement 14 600 tonnes and KG 9.6 m. She loads cargo as follows:

Weight (tonnes)	KG (m)
2500	4.5
1600	12.5

How much cargo may she load at Kg 16.0 m if the load KG is to be 10.0 m.

3. A vessel has displacement 16 000 tonnes KG 9 m. She loads the following cargo:

Wieght (tonnes)	kg (m)
1000	8
2000	6
1500	10

How should a further 2000 tonnes of cargo be distributed between spaces with Kg 5 m and Kg 11 m in order that the vessel should sail with a KG of 8.75 m.

4. A vessel summer load draft 10 m is floating at a draft of 9.8 m in water density 1.010 tonne/m^3 in a summer zone. The vessel has a TPC of 20 tonne/cm and a FWA 210 mm.

The vessel displacement at 9.8 m draft is 16 000 tonnes and she has *KG* 9.6 m. The vessel is to load cargo so as to sail at the maximum permissable draft.

Space is available at *KG* 12 m and *KG* 8 m. Distribute remaining cargo so that the *KG* of the vessel does not exceed 9.65 m.

KG (ANSWERS)

1. 8.483 m.

2. 2598.3 tonnes.

3.   684 tonnes at *Kg* 5 m.
    1316 tonnes at *Kg* 11 m.

4. DWA 126 mm.
    Load 469 tonnes at *KG* 12 m.
         183 tonnes at *KG* 8 m.

4 Conditions of equilibrium

This chapter defines the three conditions of equilibrium for a solid body and applies those conditions to the initial stability of a floating body. The metacentre is introduced as an aid to assessing initial stability.

OBJECTIVES

1. Definition and comprehension of stable, unstable and neutral equilibrium.
2. Application of the concept of equilibrium to the initial stability of vessels.
3. Definition and determination of the position of the transverse metacentre.

A body is in equilibrium whenever the resultant of the forces acting on the body is equal and opposite to the force of gravity acting on the body and the forces are in the same vertical line.

CONDITIONS OF EQUILIBRIUM

There are three possible conditions of equilibrium for a body. These are defined in the following paragraphs.

Stable equilibrium

In *Figure 4.1* when the body mass M is disturbed from an equilibrium position a couple is formed by the force of gravity, Mg, and the reaction R. The couple tends to return the body to the equilibrium position.

Note that if the centre of the gravity G was initially at a height h_0 above the datum and in the disturbed condition a distance h_1 above the datum.

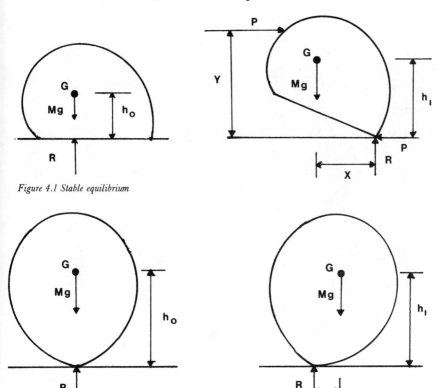

Figure 4.1 Stable equilibrium

Figure 4.2 Unstable equilibrium

Then in stable equilibrium $h_1 > h_0$, i.e. the potential energy of the body has been increased.

In general a body is in stable equilibrium if the potential energy of the body is at a minimum. The work done on the body is equal to the change in potential energy. In the context of ship stability this quantity is called the dynamical stability in the case illustrated.

$$\text{Dynamical stability} = Mg(h_1 - h_0)$$

It is also useful to see that if the disturbance was caused by a couple $P \times y$ then the body would reach another equilibrium position when

$$Py = Mgx$$

Unstable equilibrium

When the body is disturbed from the equilibrium position (see *Figure 4.2*) a couple is formed which tends to move the body away from the equilibrium position towards another position of equilibrium.

In this case $h_1 < h_0$ and the potential energy of the body has been reduced by $Mg(h_0 - h_1)$, i.e. in unstable equilibrium potential energy is at a maximum.

Neutral equilibrium

When the body is disturbed no couple is formed (*Figure 4.3*).

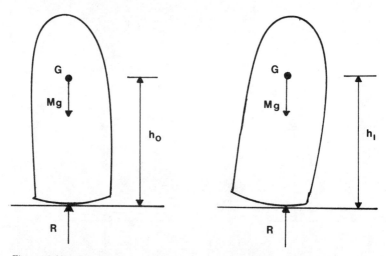

Figure 4.3 Neutral equilibrium

In this case $h_0 = h_1$, i.e. there is no change in the potential energy of the body.

STABLE EQUILIBRIUM FOR A SHIP

When a vessel which is initially upright is inclined by an external force, a couple is formed by the force of gravity acting through G with magnitude W and the force of buoyancy acting through the centre of buoyancy at B_1 with magnitude W (see *Figure 4.4*). The couple tends to return the vessel to the upright. The horizontal separation between the vectors representing is measured from G to a point Z on the vector through B_1. ($\angle Z = 90°$). The couple is called the righting moment.

$$\text{Righting moment} = W \times GZ$$

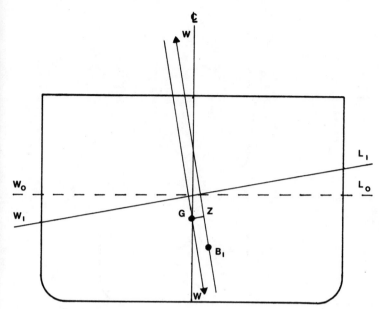

Figure 4.4 Ship in stable equilibrium

NEUTRAL EQUILIBRIUM FOR A SHIP

When a vessel which is initially upright is inclined by an external force, the force W acting through the centres of gravity G and centre of buoyancy B remain in the same vertical line (see *Figure 4.5*).

For a ship shape this condition can only be maintained for a very small angle of heel after which, in normal conditions a small righting moment tending to return the vessel to the upright would develop. Under unfavourable conditions it is possible that the vessel would become unstable and could capsize.

For a floating body to be in neutral equilibrium successive vectors through B must all intersect at G. This can only be the case for homogenius spheres, cylinders, or to a good approximation for swamped vessels. A ship can only be considered as being in neutral equilibrium if the initial shape of the vessel is regarded as being a cylinder having a diameter equal to the breadth of the vessel.

UNSTABLE EQUILIBRIUM FOR A SHIP (*Figure 4.6*)

When a vessel which is initially upright is inclined by an external force, the couple formed by the force W acting through G and B_1 will tend to heel the vessel further (*Figure 4.6*).

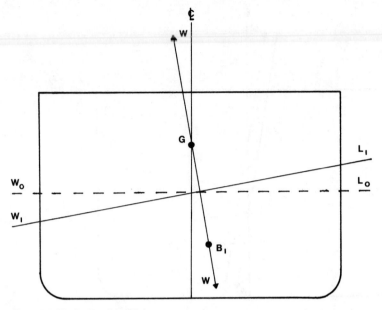

Figure 4.5 Ship in neutral equilibrium

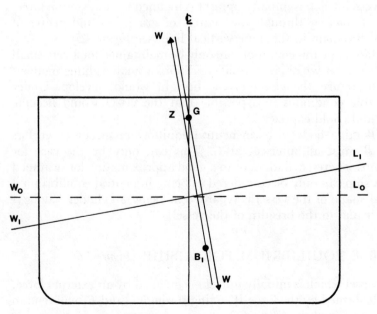

Figure 4.6 Ship in unstable equilibrium

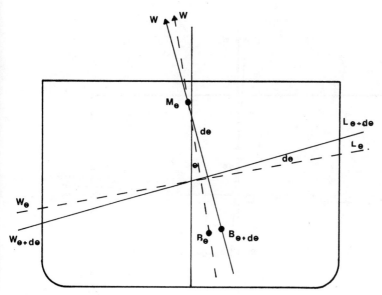

Figure 4.7 Transverse upright metacentre heeled

A vessel initially in this condition will not necessarily capsize. She will often come to rest at angle of heel called the angle of loll. In theory this angle of loll may be to port or starboard. In practice this will only be the case if G is exactly on the centreline, which would be most unlikely. The vessel will be stable relative to the angle of loll, i.e. at the angle of loll the forces acting through G and B will not form a couple and if disturbed she will tend to return to the angle of loll.

A more detailed consideration of unstable equilibrium is given in Chapter 6.

ASSESSMENT OF INITIAL STABILITY

The centre of buoyancy, at the centre of volume of the immersed part of the vessel, can be found easily for the upright condition. It is tedious to find the position of the centre of buoyancy for each angle of heel for a particular displacement.

However, we can define a point called the metacentre (M) as the intersection of successive buoyancy vectors when the vessel is heeled through a small angle $d\theta$ (*Figure 4.7*). Then if the vessel is initially upright, the initial position of the metacentre must be on the centreline (*Figure 4.8*).

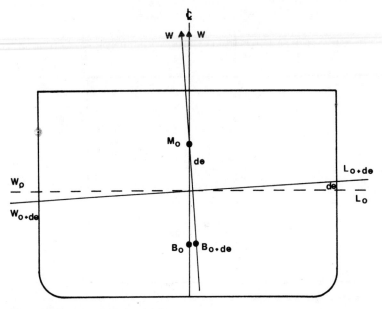

Figure 4.8 Transverse metacentre upright

Further successive points of intersection as the vessel is heeled through small angles will continue to intersect approximately at the centreline. This will continue for as long as it can be assumed that the vertical sides of the main body of the vessel are an approximation to the sides of a cylinder diameter B.

For a vessel which is initially stable M_0 will be above G. The vessel has a positive GM.

For a vessel which is initially unstable M_0 will be below G. The vessel has a negative GM.

For a vessel which is in neutral equilibrium M_0 will coincide with G. The vessel will have zero GM.

The reason for using M_0 is that it is not difficult to calculate values for BM_0 and if KB is known find $KM_0 = (KM_0 KG + BM_0)$. Then if KG is known a quick assessment of initial stability can be made.

TRANSVERSE METACENTRE OF A BOX SHAPED VESSEL

Refer to *Figures 4.9(a) and (b)*. If a vessel has length L breadth B draft d and is floating at waterline $W_0 L_0$. The vessel is now inclined through a small angle θ to waterline $W_1 L_1$. A wedge of buoyancy will move across the vessel, the centroid of the wedge will move from g_0 to g_1.

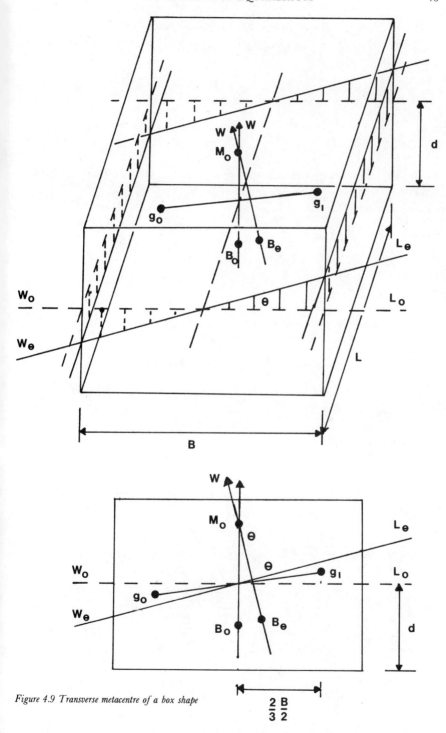

Figure 4.9 Transverse metacentre of a box shape

If the cross section in *Figure 4.9(b)* is now considered at g_0g_1.

If the volume of the wedge of buoyancy is v and the underwater volume is ∇ then

$$B_0B_1 = \frac{v \times g_0g_1}{\nabla}$$

and

$$B_0B_1 = B_0M\theta_c$$

(as θ_c, in circular measure, is approximately equal to tan θ and sin θ if θ is small).

Now $g_0g_1 = 2 \times \dfrac{2}{3} \times \dfrac{B}{2}$

$$= \frac{2}{3} B$$

$$v = \frac{1}{2} \times \frac{B}{2} \times \frac{B}{2} \times \theta_c \times L$$

$$= \frac{LB^2}{8} \theta_c$$

$$v \times g_0g_1 = \frac{2}{3} B \times \frac{LB^2}{8} \theta_c$$

$$= \frac{LB^3}{12} \theta_c$$

(Note that $LB^3/12$ is the second moment of area of a rectangle about the axis through the centroid)

$$B_0B_1 = \frac{LB^3}{12} \frac{1}{LBd} \theta_c$$

$$B_0B_1 = \frac{B^2}{12d} \theta_c$$

$$B_0M\theta_c = \frac{B^2}{12d} \theta_c$$

$$B_0M = \frac{B^2}{12d}$$

$$KB = \frac{d}{2}$$

$$KM = \frac{d}{2} + \frac{B^2}{12d}$$

(Note that this could be written as

$$KM = KB + \frac{I}{\nabla}$$

where I is the second moment of area of the waterplane about the centreline.)

Considering the equation for a box shape

$$KM = KB + BM$$

$$KB = \frac{d}{2} + \frac{B^2}{12d}$$

it can be seen that when draft (d) is small BM will be large and that as draft increases KB will increase, however BM will steadily reduce as draft increases. If draft could be made very large

$$KM = \frac{d}{2} = KB$$

Example 4.1 (*Figure 4.10*)

A box shaped vessel has length, 100 m and breadth, 12 m and floats at a range of drafts from 1 m to 10 m.

Produce curves of KB_1, BM and KM (*Figure 4.10*)

d	1	2	3	4	5	6	7	8	9	10
$KB \frac{d}{2}$	0.5	1.0	1.5	2.0	2.5	3.0	3.5	4.0	4.5	5.0
$BM \frac{12}{d}$	12.0	6.0	4.0	3.0	2.4	2.0	1.72	1.5	1.33	1.2
KM	12.5	7.0	5.5	5.0	4.9	5.0	5.22	5.50	5.83	6.2

Thus it can be seen that KM reaches a minimum value. Differentiating

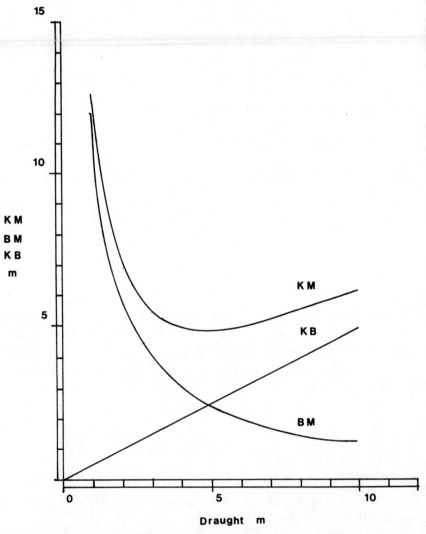

Figure 4.10 Curves of KB, BM, KM for a box shape

with respect to draft

$$\frac{d(KM)}{d(d)} = \frac{1}{2} - \frac{B^2}{12d^2}$$

Equating to zero to find minimum

$$0 = \frac{1}{2} - \frac{B^2}{12d^2}$$

$$d^2 = \frac{13^2}{6}$$

$$d = \frac{B}{\sqrt{6}}$$

Putting $B = d\sqrt{6}$

$$KM = \frac{d}{2} + \frac{(d\sqrt{6})^2}{12d}$$

$$KM = \frac{d}{2} + \frac{6d^2}{12d} = d$$

Thus for a box shape KM is a minimum when

$$d = \frac{B}{\sqrt{6}}$$

and at this draft

$$KM = d$$

Example 4.2

A box shaped vessel length, 200 m, breadth 20 m and depth 10 m is loades to that KG of the vessel is always equal to the draft.

Find the maximum draft at which the vessel will be stable, and the GM at minimum KM.

If the vessel had draft d

$$KG = d$$

$$KM = KB + KM$$

For a box shape

$$KM = \frac{d}{2} + \frac{B^2}{12d}$$

$$KM = \frac{d}{2} + \frac{400}{12d}$$

$$KM = \frac{d}{2} + \frac{33.33}{d}$$

For

$$GM = 0$$

$$KG = KM$$

$$d = \frac{d}{2} + \frac{33.33}{d}$$

$$2d^2 = d^2 + 66.67$$

$$d^2 = 66.67 \text{ m}$$

$$d = 8.165 \text{ m}$$

By inspection of typical KM curve, vessel will be stable up to 8.165 m draft and unstable at greater drafts.

Example 4.3

A floating body has square cross section of side 1 m and KG always equal to 0.5 m. Find the range of relative densities over which the body will be (a) stable and (b) unstable, when floating in fresh water. What will be the minimum GM and at which draft does it occur.

$$KG = 0.5 \text{ m}$$

$$KM = \frac{d}{2} + \frac{B^2}{12d}$$

$$KM = \frac{d}{2} + \frac{1}{12d}$$

For vessel to be stable $KM > KG$. Putting

$$KG = KM$$

$$0.5 = \frac{d}{2} + \frac{1}{12d}$$

$$6d = 6d^2 + 1$$

$$0 = 6d^2 - 6d + 1$$

$$d = 0.789 \text{ m or } 0.211 \text{ m}$$

By inspection of box-shaped KM curve, vessel will be stable for

$$RD < 0.211 \text{ and } RD > 0.789$$

Vessel will be unstable for

$$0.211 \leqslant RD \leqslant 0.789$$

$K\bar{M}$ minimum occurs at

$$d = \frac{B}{\sqrt{6}} = \frac{1}{\sqrt{6}} = 0.408 \text{ m}$$

at this draft

$$KM = d$$

$$\begin{array}{rl} = & 0.408 \\ KG = & 0.500 \\ \hline GM = & -0.092 \\ \hline \end{array}$$

BM FOR A SHIP SHAPE

If we consider a wedge of buoyancy moving across the vessel as in the work in box shapes we have as before in *Figure 4.12*

$$B_0 B_1 = \frac{v \times g_0 g_1}{\nabla}$$

If we now consider an elementary segment of the length of the vessel dl (*Figure 4.11*). If the volume of the elementary wedge is S and the centroid of the element moves from b_0 to b_1.

For the element

$$b_0 b_1 = 2 \times \frac{2}{3} \times \frac{y}{2} = \frac{2}{3} y$$

$$S = \frac{1}{2} \frac{y}{2} \times \frac{y}{2} \theta_c \, dl = \frac{y^2}{8} \theta_c \, dl$$

$$S \times b_0 b_1 = \frac{2}{3} y \times \frac{y^2}{8} \theta_c \, dl = \frac{y^3 \, dl}{12}$$

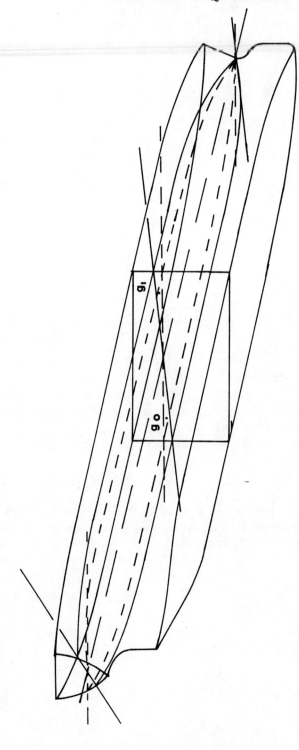

Figure 4.11 Transverse metacentre of a ship shape

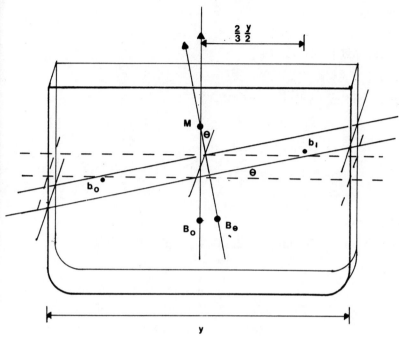

Figure 4.12 Element of length of ship shape

For the entire wedge

$$V \times g_0 g_1 = \theta_c \int_0^L \frac{y^3}{12} \, dl$$

but $\displaystyle\int_0^L \frac{y^3}{12} \, dl$

is I the second moment of area of the water plane (Chapter 1).
  Also for the entire vessel

$$B_0 M \theta_c = B_0 B_1$$

$$B_0 M \theta_c = \frac{I}{V} \, \theta_c$$

$$B_0 M = \frac{I}{V}$$

KM and BM can be determined using the numerical methods described
in Chapter 1.

The relationship

$$KB = KB + BM = KB + \frac{I}{\nabla}$$

can be extended to all floating shapes.

Example 4.4

A vessel has the following $\frac{1}{2}$ areas of waterplane at the drafts given:

Draft	0.25	0.75	1.25	2.25	3.25	4.25	5.25 m
$\frac{1}{2}$ area	800	1600	2300	2600	2750	2800	2825 m^2

At draft 5.25 m the vessel has the following $\frac{1}{2}$ ordinates of waterplane commencing from aft.

Station	0	1	2	3	4	5	6	7	8
$\frac{1}{2}$ Ord.	0	12.0	14.5	15.0	15.0	14.0	12.0	8.0	3.5 m

Below the 0.25 m draft there is an appendage volume 150 m^3 Kb 0.2 m.
    Forward of station 8 there is an appendage second moment of area 30 m^4.
    Length of the vessel 240 m from station 0 to 8
KB

b 0.25d to 5.25 m

Draft (m)	$\frac{1}{2}$ Area (m^2)	Simpson's multiplier	Function Vol.	Lever	Function First moment
0.25	800	$\frac{1}{2}$	400	0	0
0.75	1600	$\frac{1}{2}$	3200	0.5	1600
1.25	2300	$1\frac{1}{2}$	3450	1	3450
2.25	2600	4	10400	2	20800
3.25	2750	2	5500	3	16500
4.25	2800	4	11200	4	22800
5.25	2825	1	2825	5	14125
			36975		79275

$$\text{Volume} = 2 \times \frac{h}{3} \times \sum \text{function volume}$$

$$= 2 \times \frac{1}{3} \times 36975 \text{ m}^3$$

$$= 24650 \text{ m}^3$$

$$0.25 - b = \frac{h \sum \text{functions first moment}}{\sum \text{function volume}}$$

$$= \frac{1 \times 79275}{36975} = 2.144 \text{ m}$$

$$Kb = (2.144 + 0.25) \text{ m} = 2.394 \text{ m}$$

Volume	Kb	Moment
24650	2.394	59012.1
150	0.200	30.0
24800		59042.1

$$KB = \frac{\text{moment of volume}}{\text{volume}}$$

$$= \frac{59042.1}{24800} \text{ m} = 2.381 \text{ m}$$

Second moment of waterplane

Station	$\frac{1}{2}$ Ord	$\frac{1}{2}$ Ord^3	Simpson's multiplier	F (Second moment)
0	0.0	0.0	1	0.0
1	12.0	1728.0	4	6912.0
2	14.5	3048.6	2	6097.2
3	15.0	3375.0	4	13500.0
4	15.0	3375.0	2	6750.0
5	14.0	2744.0	4	10976.0
6	12.0	1728.0	2	3456.0
7	8.0	512.0	4	2048.0
8	3.5	42.9	1	42.9
				49782.1

$$I_{CL} = \left(2 \times \frac{h}{3} \times \frac{1}{3} \times \sum F \text{ Second moment} + 30\right) \text{ m}^4$$

$$= 2 \times \frac{30}{3} \times \frac{1}{3} \times 49782.1 + 30 = 331910.7 \text{ m}^4$$

$$BM = \frac{I}{\nabla}$$

$$= \frac{331910.7}{24800}$$

$$= 13.38 \text{ m}$$

$$KB = 2.39 \text{ m}$$

$$KM = 15.77 \text{ m}$$

Example 4.5

A vessel displacement 22 600 tonnes, KG 8.2 m discharges 3000 tonnes of ballast from a mean Kg of 2.0 m.

She loads 11 800 tonnes of cargo at a mean Kg of 7.8 m. A further parcel of 1200 tonnes of cargo remains to be loaded.

Determine the mean Kg at which to load this cargo so that the final GM is at least 0.5 m.

KM at displacement 32 200 tonnes is 9.0 m.

Let x be the mean Kg at which to load cargo.

Weight	Kg	Moment
22600	8.2	185300
11400	7.8	88900
1200	x	1200x
−3000		−6000
32200		268200 + 1200x

$$\text{Max. } KG = KM - GM$$
$$= (9.0 - 0.5) \text{ m} = 8.5 \text{ m}$$

$$KG = \frac{\text{moment of weight}}{\text{displacement}}$$

$$8.5 = \frac{268200 + 1200x}{32200}$$

$$x = 4.67 \text{ m}$$

Example 4.6 (*Figure 4.13*)

Making use of the hydrostatic curves for *MV Nonesuch*. *MV Nonesuch* is floating at draft 6.20 m and *KG*, 10.3 m.

She is to load cargo as follows:

No. 1   3000 tonnes   *KG* 9.0 m
No. 3   2500 tonnes   *KG* 8.0 m
No. 5   2800 tonnes   *KG* 8.6 m

Distribute cargo between No. 2 hold at *Kg* 3 m and on deck at *Kg* 18 m so that the final draft is 9.0 m and the *GM* is no less than 1.4 m.

At draft 6.20 m, displacement 21 600 tonnes.

Weight	Kg	Moment
21600	10.3	222480
3000	9.0	27000
2500	8.0	20000
2800	8.6	24080
——		——
29900		293560

At draft 9.0 m, displacement   31000 tonnes
Present displacement   29900 tonnes

Cargo to load   1100 tonnes

At draft 9.0 m, *KM*   11.4 m
Min *GM*   1.4 m

Max *KG*   10.0 m

Let w be cargo to load at *Kg*, 18 m

Weight	Kg	Moment
29900		293560
w	18	$+18w$
$1100-w$	3	$3300-3w$
——		——
31000		$296860+15w$

$$KG = \frac{\text{moment of weight}}{\text{weight}}$$

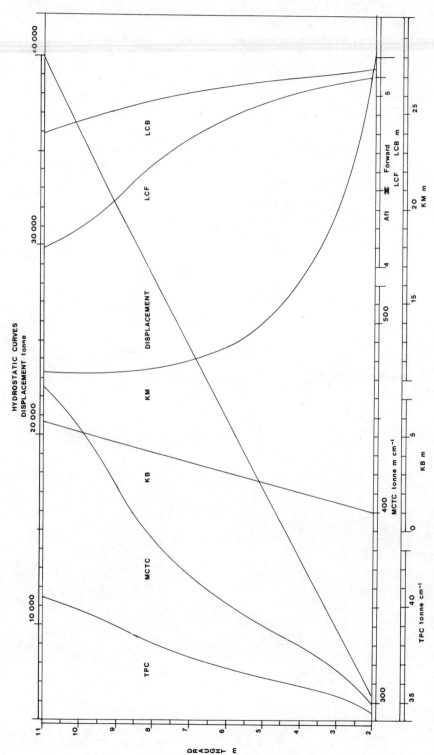

Figure 4.13 Hydrostatic curves—MV Nonesuch

$$10 = \frac{296860 + 15w}{31000}$$

$$310000 = 296860 + 15w$$

$$15w = 13140$$

$$w = \quad 876 \text{ tonnes}$$

Load 876 tonnes at KG 18 m
224 tonnes at KG 3 m

QUESTIONS ON GM

1. A vessel displacing 8000 tonnes has KG 7.4 m. She is to make a passage during which she will use 200 tonnes of fuel from kg 0.5 m and 100 tonnes of FW from Kg 1.5 m. She is to load timber on deck at a Kg of 12 m.

   It is anticipated that by the time the vessel arrives at her destination the deck cargo will have absorbed 20% of its own weight of water. How much timber may she load if her arrival GM is to be 0.4 m.

   KM may be assumed to be 8.2 m over the range of displacements being considered.

2. A vessel has displacement 10 900 tonnes, KG 7.0 m.

   Distribute 5742 tonnes of cargo between spaces Kg 8.17 m at Kg 7.43 m so that the final GM is 1.24 m.

   KM at displacement 16 642 tonnes 8.43 m.

3. A box-shaped vessel length 100 m, breadth 10 m, depth 5 m displacing 2000 tonnes has KG 4.5 m and is floating in fresh water.

   Find the initial GM of the barge and the GM after a weight of 500 tonnes is loaded at KG 4.0 m. Find the righting moment at $10°$ heel in both cases.

4. A vessel displacing 10 900 tonnes has KG, 6.20 m, KM, 7.20 m. A heavy lift weighing 200 tonnes is onboard at Kg 2.6 m.

   Calculate how much ballast must be loaded at Kg 1.0 m if the GM is to be maintained at GM 1.0 m, after the heavy lift has been discharged. KM may be assumed to remain constant.

5. A vessel has the following half areas of waterplane at the drafts given:

Draft	0.2	0.6	1.0	1.8	2.6	3.4	4.2 m
Area	200	950	1490	1705	1825	1850	1861 m^2

At draft 4.2 m the vessel has the following half ordinates of waterplane at equal intervals commencing from aft.

Station	0	1	2	3	4	5	6	7	8	9
Half Ord.	0.0	1.9	5.2	9.3	12.4	13.1	13.1	12.0	6.2	1.1 m

Forward of station 9 there is an area of 3 m^2
Second moment of area, 6 m^4
Length of the vessel, 225 m.
Below draft 0.2 m there is an appendage volume 130 m^3 Kb, 0.16 m.
Find the KM of the vessel.

6. MV Nonesuch is floating at a draft 5.5 m, KG 11.0 m. She is to load cargo as follows:

	Weight (tonnes)	KG (m)
No. 1	3800	7.6
No. 2	2800	8.3
No. 3	2500	5.6
No. 4	3100	6.4
No. 5	3200	6.8

How much cargo can she load into No. 1 TSWT Kg, 15 m and No. 4 hold Kg, 8.0 m if the maximum draft is to be 10.0 m and the GM no less than 2.0 m.

GM (ANSWERS)

1. Load, 220.2 tonnes of timber.

2. Load, 4805.5 tonnes at Kg 7.43 m.
          936.5 tonnes at Kg 8.17 m.

3. Initial GM, 0.67 m.
    Righting moment, 232.7 tonnes.
    Final GM, 0.18 m.
    Righting moment, 78.14 tonnes m.

4. Load, 138.4 tonnes of ballast.

5. Volume 0.2 m–4.2 m   12 781.8 m^3
    Kb   0.2 m–4.2 m   2.426 m
    ∇   12 912 m^3
    KB   155 704.9 m^4
    BM   12.058 m
    KM   14.461 m

6. Load, 1725 tonnes
            550.2 tonnes at *Kg* 15.0 m
            1174.8 tonnes at *Kg* 8.0 ,.

5 List and free surface

This chapter considers the effect of the centre of gravity of the vessel not being on the centreline, the effect of having liquids within the vessel which are free to move and the effect of suspending weights.

OBJECTIVES

1. To calculate the list of a vessel at small angles of list.
2. To assess the effect on the initial stability a vessel of having liquid within the vessel which is free to move.
3. To assess the effect on the initial stability of a vessel when weights are suspended.
4. To find the position of the centre of gravity of the vessel using the inclining test.

LIST

If the centre of gravity of the vessel is not on the centreline the vessel will incline until the centre of gravity and centre of buoyancy are once more in the same vertical line.

Provided the list is small it can be assumed that the vertical line through G and B intersects the centreline at the metacentre (M). If the angle of list is large then the list has to be determined using GZ curves (Chapter 6).

In *Figure 5.1(a)* a vessel has displacement W with centre of gravity at G_0, on the centreline, centre of buoyancy at B_0 with transverse metacentre at M. w is a small weight already onboard.

The weight w is moved a distance dm transversely across the vessel (*Figure 5.1(b)*). The centre of gravity will move from G_0 to G_1. There will be a heeling moment $W \times G_0G_1$.

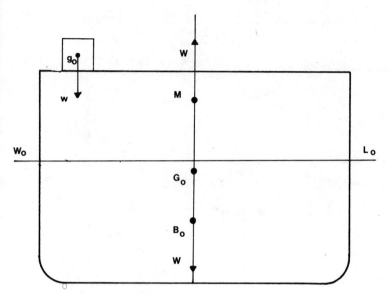

Figure 5.1(a) List initial condition

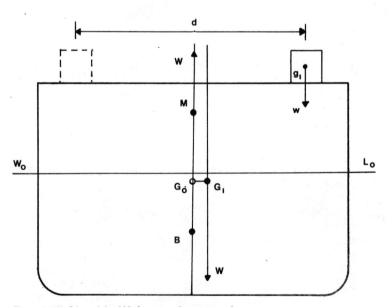

Figure 5.1(b) List weight shifted transversely across vessel

In *Figure 5.1(c)* the vessel inclines until G_1 and B_1 are in the same vertical line. Then provided the angle of list θ is small, this vertical line will pass through M the transverse metacentre.

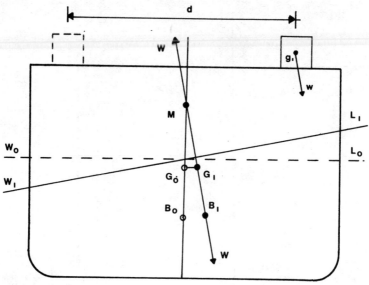

Figure 5.1(c) List final condition

Angle $\quad MG_0G_1 = 90°\quad \therefore \quad\quad\quad \tan\theta = \dfrac{G_0G_1}{G_0M}$

In the particular case of the single weight moving across the vessel

$$G_0G_1 = \frac{w \times d}{W}$$

$$\tan\theta = \frac{w \times d}{WG_0M}$$

A more practical approach is to consider weights as being loaded at various Kg away from the centreline.

Example 5.1

A vessel displacing 11 000 tonnes KG, 8.7 m, KM 9.5 m has an initial list of 2° to port.

She loads	Weight (tonne)	Kg (m)	Distance from centreline (m)
	400	10.0	4.5 port
	600	4.0	6.0 starboard
and discharges	100	1.0	2.0 port

Find the final list

$$\tan \theta = \frac{G_0 G_1}{G_0 M} \qquad\qquad G_0 M = KM - KG$$

$$\begin{aligned} G_0 G_1 &= G_0 M \tan \theta & &= 9.5 - 8.7 \\ &= 0.8 \tan 2^\circ & &= 0.8 \text{ m} \\ &= 0.0279 \text{ m port} \end{aligned}$$

Weight (tonne)	KG (m)	Moment (tonne m)	G from centreline (m)	Moment port (tonne m)	Moment starboard (tonne m)
11 000	8.7	95 700	0.0275 p	306.9	—
400	10.0	4000	4.500 p	1800.0	—
600	4.0	2400	6.000 s	—	3600
− 100	1.0	− 100	2.000 p	− 200.0	—
11 900		102 000		1906.9	3600
					− 1907
					1693 s

$$KG_0 = \frac{102\,000}{11\,900} \qquad\qquad G_0 G_1 = \frac{\text{moment about centreline}}{\text{displacement}}$$

$$= 8.571 \text{ m} \qquad\qquad\qquad = \frac{1693}{11\,900}$$

$$KM = 9.500 \text{ m} \qquad\qquad\qquad = 0.142 \text{ m}$$

$$= 0.929 \text{ m} \qquad\qquad \tan \theta = \frac{G_0 G_1}{G_0 M}$$

$$= \frac{0.142}{0.929}$$

$$\theta = 8.71^\circ = 8^\circ\, 42' \text{ starboard}$$

EFFECT OF SUSPENDING WEIGHTS (*Figure 5.2*)

A vessel with displacement W and centre of gravity at G_0 has on board a weight w. If the weight is now suspended from a point O, a distance d_1 above the weight and the vessel is inclined by an external force to a small angle θ. The weight will move through a distance d_2.

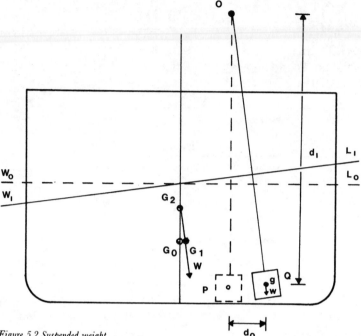

Figure 5.2 Suspended weight

In triangles $G_1 G_0 G_2$, OPQ

$$\frac{G_1 G_0}{G_0 G_2} = \frac{OP}{PQ} = \frac{d_1}{d_0}$$

$$G_0 G_2 = \frac{w \times d_0}{W}$$

$$\frac{G_0 G_1 \times W}{w \times d_0} = \frac{d_1}{d_0}$$

$$G_0 G_1 = \frac{w \times d_1}{W}$$

i.e. there is an effective rise of the centre of gravity from G_0 to G_1 which is the distance the centre of gravity would have moved, had w been moved to O. The vessel responds as if the suspended weight had its centre of gravity at the point of suspension.

Thus if a heavy lift is picked up from a hold and swung overside, the centre of gravity of the vessel firstly moves vertically upwards, before moving away from the centreline as the lift is moved away from the centreline.

Example 5.2

A vessel displacement 18 000 tonnes is upright and has KM, 9.7 m. A heavy lift is onboard at Kg, 10.9 m, 7 m to port of the centreline, mass 90 tonnes.

The lift is to be discharged into a lighter, at which time the derrick will be plumbed 12 m to starboard of the centreline and the head of the derrick will be 29 m above the keel.

What must the KG of the vessel be if the list is not to exceed 7°?

Let the KG of the vessel be x.

Weight	KG	Moment	G from centreline	Moment P	Moment S
18 000	x	18 000x	0	—	—
90	29	2610	12s	—	1080
−90	10.9	−981	7p	−630	—
18 000		1629 + 18 000x		−630	1080 −(−630)
					1710

$$KG = \frac{1629 + 18\,000x}{18\,000}$$

$$G_0M = 9.7 - \frac{1629 + 18\,000x}{18\,000}$$

$$= \frac{9.7 \times 18\,000 - 1629 - 18\,000x}{18\,000}$$

$$G_0G_1 = \frac{1710}{18\,000}$$

$$G_0M = \frac{172\,971 - 18\,000x}{18\,000}$$

$$= 0.095 \text{ m}$$

$$\tan 7° = \frac{G_0G_1}{G_0M} = \frac{0.095 \times 18\,000}{172\,971 - 18\,000x}$$

$$0.1229 = \frac{0.095 \times 18\,000}{172\,971 - 18\,000x}$$

$$0.1229(172\,971 - 18\,000x) = 1710$$

$$21\,258 - 2212.2x = 1710$$

$$-2212.2x = -19\,548$$

$$x = 8.836 \text{ m}$$

A vessel will be upright when

Moments to ports of centre
= moments to starboard of centreline

Example 5.3

A vessel displacing 11 500 tonnes has KG, 7.5 m; KM, 8.4 m and is listed 4° to port. The vessel has yet to load 750 tonnes.

Space is available at Kg 10.5 m, 6 m to port of the centreline and at Kg 8.0 m, 4 m to starboard of the centreline.

Distribute the cargo so that the vessel completes upright and find the final KG.

Load w tonnes at Kg 10.5 m

$$G_0G_1 = G_0M \tan \theta$$
$$= 0.9 \tan 4° \text{ port}$$
$$= 0.0629 \text{ m}$$

Weight	KG	Moment	G from centreline	Moment P	Moment S
11 500	7.5	86 250	0.0629p	723.4	
w	10.5	10.5w	6.0000	6.0x	
750 − w	8.0	6000 − 8.0w	4.000	—	3000.0 − 4w
12 250		92 250 + 2.5w		723.4 + 6w	3000.0 − 4w
					723.4 + 6w
					2276.6 − 10w

$$KG = \frac{92\,250 + 2.5w}{12\,250}$$ For ship to be upright

$$KG = \frac{92\,250 + 569.3}{12\,250}$$ $2276.6 - 10w = 0$

$$KG = 7.577 \text{ m}$$ $w = 227.7 \text{ tonnes}$

Load 227.7 tonnes at Kg 10.5 m
522.3 tonnes at Kg 8.0 m

QUESTIONS ON LIST

1. A vessel, displacement 10 500 tonnes, is floating upright KG, 7.8 m; KM, 8.5 m. Caro weighin 300 tonnes is loaded at Kg, 10 m; 5 m to port of the centreline. Find the final list.

2. A vessel displacement 8450 tonnes is listed 6° to starboard, KG, 7.8 m; KM, 8.5 m. She loads 250 tonnes of ballast at Kg, 1.5 m; 3.1 m to port of the centreline. Find the final list.

3. A vessel has displacement 12 000 tonnes, Kg, 8.4 m tonnes, and is listed 5° to starboard. A total of 600 tonnes of cargo is to be loaded at a Kg of 10 m.

    If the cargo is to be loaded in compartments centres of gravity 6 m to port and 5 m to starboard of the centreline. Distribute the cargo so that the vessel completes upright. KM of vessel, 9.0 m.

4. A vessel has displacement 15 500 tonnes and has KM 10.5 m and is upright. She is to load a heavy lift weighing 100 tonnes and the list is not to exceed 5° at any time. The lift is to be loaded using a derrick the head of which is 30 m above the keel and will be plumbed a maximum of 12 m to starboard of the centreline.

LIST (ANSWERS)

1. Final KG, 7.861 m. List, 9.85° port.

2. Final KG, 7.619 m. List, 1.13° port.

3. Final KG load, 270 tonne to starboard. 330 tonne to port.

4. Maximum initial KG, 9.49 m.

FREE SURFACES

So far, the effect of adding removing shifting and suspending weights has to be examined. There is one other case, that is where liquid is free to move continuously as a result of the vessel being heeled by an external force.

In *Figure 5.3(a)* a vessel has on board liquid weight w with centre of gravity at g_0, the liquid does not fill the tank. The vessel had displacement W with centre of gravity at G_0, centre of buoyancy at B_0 and metacentre at M. If the vessel is heeled to waterline W_1L_1, (*Figure 5.3(b)*) the centre of gravity of the liquid in the tank will move to g_1, causing the centre of gravity of the vessel to move to G_1. Now at this angle of heel the centre of buoyancy has moved B_1, the righting lever has been reduced from G_0Z_0 to G_1Z_1 (*Figure 5.3(c)*).

The movement of the centre of gravity is similar to the movement of the centre of buoyancy as the vessel heels. Hence it can be assumed for small angles of heel that verticals through successive positions of G intersect the centreline at a fixed point G_v. This point is called the virtual centre of gravity or the fluid centre of gravity.

For small angles of heel the vessel behaves as if the centre of gravity of the vessel was at G_v. From now on it is important to distinguish between KG_{solid} and KG_{fluid} when defining the stability of a vessel. The assumption

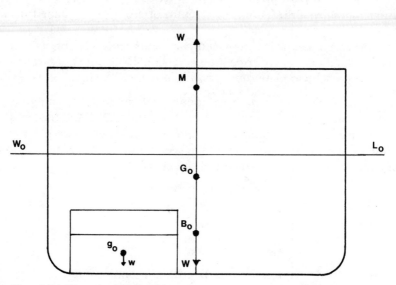

Figure 5.3(a) *Free surface initial condition*

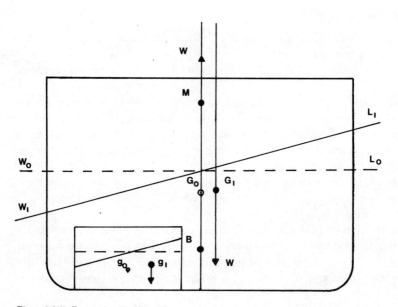

Figure 5.3(b) *Free surface liquid moved*

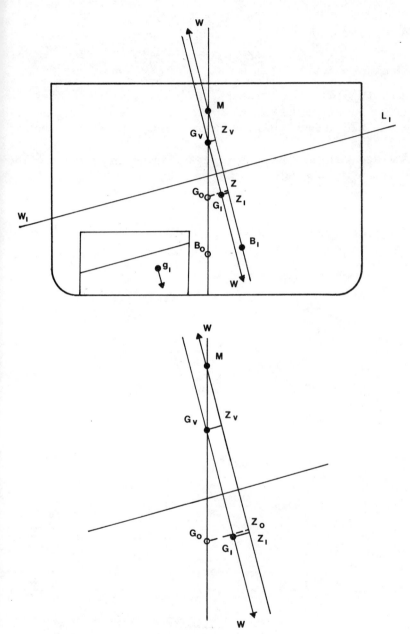

Figure 5.3(c) Free surface final condition

that G_v is a fixed point is reasonable so long as the actual transfer of liquid can be regarded as vertically sided wedges.

When there is free surface on a vessel GZ is always reduced.

Evaluation of G_0G_v

On board a vessel displacement W with centre of gravity at G_0 there is a rectangular tank length l and breadth b containing liquid density ρ_i. (See *Figure 5.4*.) The vessel is inclined by an external force to some angle of heel θ.

A wedge of liquid moves across the tank, the centre of gravity of the liquid moves from g_0 to g_1. Then if the wedge of liquid has weight w

$$G_0G_1 = \frac{w \times g_0g_1}{W}$$

$$w = v \times \rho_i$$

where v is the volume of the wedge

$$= \frac{1}{2} \times \frac{b}{2} \times \frac{b}{2} \theta_c \times \rho_i \times l$$

$$= \frac{l \times b^2}{8} \rho_i \theta_c$$

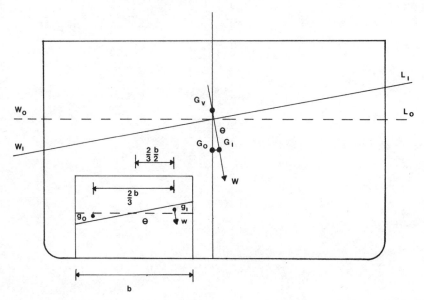

Figure 5.4 Virtual rise in G due to free surface

$$g_0 g_1 = 2 \times \frac{2}{3} \times \frac{b}{2}$$

$$= \frac{2}{3} \times b$$

$$w \times g_0 g_1 = \frac{l \times b^2}{8} \times \rho_i \times \theta_c \times \frac{2}{3} \times b$$

$$= \frac{l \times b^3}{12} \times \rho_i \times \theta_c$$

$$\frac{l \times b^3}{12} = \text{the second moment of area of a rectangle about}$$
$$\text{a longitudinal axis through the centroid}$$

$$= i$$

$$G_0 G_1 = \frac{i \times \rho_i \times \theta_c}{W}$$

But

$$G_0 G_1 = G_0 G_v \times \theta_c$$

$$G_0 G_v \theta_c = \frac{i \times \rho_i \times \theta_c}{W}$$

$$G_0 G_v = \frac{i \rho_i}{W}$$

If the tank is not rectangular and we consider an elementary length of the tank

$$v \times g_0 g_1 = \theta \rho_i \int_0^l \frac{b^3}{12} \, dl$$

$$= i \rho_i \theta_c \text{ as above}$$

$$G_0 G_v = \frac{i \rho_i}{W} \quad \text{for all tank shapes}$$

Note that for a particular tank i is a constant, for a particular depth of liquid, and has units of m^4. If we take the product $i \rho_i$ the units are $m^4 \times$ tonne/m^3 = tonne m, a moment. If ρ_i is taken as being the density of fresh water then $i \rho_{FW}$ is called the Free Surface Moment (FSM) for fresh water.

Then if some other liquids are carried.

$$\text{free surface moment} = \text{free surface moment}_{FW} \times RD$$

The value of free surface moment can be used either as illustrated above, i.e.

$$G_0G_v = \frac{\text{FSM}}{W}$$

or the value for FSM can be added to the moment column in the standard 'Weight KG Moment' calculation.

EFFECT OF LONGITUDINAL SUBDIVISION (*Figure 5.5*)

If a rectangular breadth b is divided into n equal subdivisions each subdivision containing liquid density ρ_i.
   Then for one tank

$$G_0G_v = \frac{i\rho_i}{W}$$

$$= \frac{l\left(\dfrac{b}{n}\right)^3 \rho_i}{12W}$$

$$= \frac{lb^3\rho_i}{12W} \times \frac{1}{n^3}$$

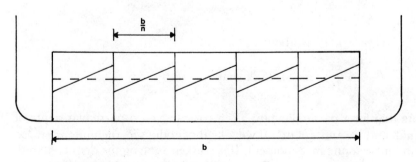

Figure 5.5 Effect of longitudinal subdivision

Since there are n tanks

$$G_0 G_v = n \frac{l b^3 \rho_i}{12W} \times \frac{1}{n^3}$$

$$= \frac{l b^3 \rho_i}{12W} \times \frac{1}{n^2}$$

Thus the virtual rise in G due to free surface is reduced by a factor of $1/n^2$ where n is the number of subdivisions.

Note that the formula can only be applied to equal rectangular tanks. If the tanks are rectangular but unequal, then the free surface of each tank must be found separately. If the tanks have the same shape and area but are not rectangular then the formula is a reasonable approximation but is not strictly accurate and for exact calculations the free surface of each tank should be considered individually.

EFFECT OF TRANSVERSE SUBDIVISION

Transverse bulkheads have no direct effect on free surface, however, they do give the operator greater flexibility in filling and emptying tanks.

In situations where large free surfaces could develop as a result of damage, transverse subdivisions would reduce the extent of the free surface.

EFFECT OF DEPTH OF LIQUID

(a) *On initial stability*
In most practical cases the formula given covers the worst condition, i.e. the maximum likely free surface. If the liquid level is low in a tank then it will flow to one side and form a pocket of liquid with considerably reduced area. Conversely if the tank is full the liquid would reach the top of the tank and again the area of the liquid would be reduced.

(b) *At large angles of heel*
When considering large angles of heel the actual position of the centre of gravity of liquid must be found. This process is tedious, but developments in computation make this practical and some stability calculation packages find the transverse position of the centre of gravity, treating liquid in tanks as weights moving transversely across the vessel.

Example 5.4

A vessel displacing 10 000 tonnes KG, 8.9 m; KM, 9.4 m. The vessel loads ballast water RD 1.010 into a rectangular tank length, 30 m; breadth, 20 m; depth, 2 m. The tank has a single centreline division to a depth of 1.0 m. Kg of ballast, 0.5 m. Find the fluid GM of the vessel.

$$\text{Weight of ballast } l \times b \times d \times \rho$$
$$= 30 \times 20 \times 1 \times 1.010 \text{ tonne}$$
$$= 606 \text{ tonne}$$

Weight	KG	Moment
10 000	8.9	89 000
606	0.5	303
10 606		89 303

$$KG \text{ solid} = \frac{\text{moment}}{\text{weight}}$$

$$= \frac{89\,303}{10\,606} = 8.42 \text{ m}$$

$$G_0 G_v = \frac{i\rho}{W} \times \frac{1}{n^2}$$

$$= \frac{lb^3}{12} \frac{\rho}{W} \frac{1}{n^2}$$

$$= \frac{30 \times 20^3}{12} \frac{1.010}{10\,606} \times \frac{1}{2^2}$$

$$GG_v = 0.476 \text{ m}$$
$$KG \text{ solid} = 8.420 \text{ m}$$

$$KG \text{ fluid} = 8.896$$
$$KM = 9.400$$

$$GM \text{ fluid} = 0.504 \text{ m}$$

Example 5.5

An undivided forward double bottom tank has the following half ordinates of water plane in a tank 18 m long.

Station	0	1	2	3	4	5	6
½ ord	2.0	2.8	3.9	5.2	6.7	8.0	10.1 m

Find the fluid KG of the vessel after loading fuel oil RD 0.9 into the tank to a depth of 0.8 m, Kg, 0.4 m. Assume tank is wall sided.
   Displacement of vessel 20 000 tonnes KG 10 m.

Station	½ ord	Simpson's mlultiplier	F (area)	(½ ord.)³	SM	F (second moment)
0	2.0	1	2.0	8.0	1	8.0
1	2.8	4	11.2	21.95	4	87.8
2	3.9	2	7.8	59.32	2	118.6
3	5.2	4	20.8	140.61	4	562.4
4	6.7	2	13.4	300.76	2	601.5
5	8.0	4	32.0	512.00	4	2048.0
6	10.1	1	10.1	1030.30	1	1030.3
			97.3			4456.7

$$\text{Area} = 2 \times \frac{h}{3} \times \sum F \text{ (area)}$$

$$= 2 \times \frac{3}{3} \times 97.3 \text{ m}^2$$

$$= 194.6 \text{ m}$$

$$\text{Weight of oil} = \text{Area} \times \text{depth} \times \rho \text{ oil}$$

$$= 194.6 \times 0.8 \times 0.9 \text{ tonne}$$

$$= 140.112 \text{ tonnes}$$

$$i = 2 \times \frac{h}{3} \times \frac{1}{3} \times \sum F \text{ (second moment)}$$

$$= 2 \times \frac{h}{3} \times \frac{1}{3} \times 4456.7 \text{ m}^4$$

$$= 2 \times \frac{3}{3} \times \frac{1}{3} \times 4456.7 \text{ m}^4$$

$$= 2971.13 \text{ m}^4$$

Free Surface Moment $= i\rho$ oil

$$= 2971.13 \times 0.9 \text{ tonne m}$$

$$= 2674.0 \text{ tonne m}$$

Item	Weight	KG	Moment
Ship	20 000	10.0	200 000
Oil	140	0.4	56
FSM			2674
	20 140		202 730

$$KG_{\text{fluid}} = \frac{202\,730}{20\,140}$$

$$= 10.066 \text{ m}$$
$$KG_0 = 10.000 \text{ m}$$

$$Diff = 0.066 \text{ m}$$

GM is reduced by 0.066 m by loading liquid in the bottom of the vessel.

Example 5.6 (*Figure 5.6*)

A vessel displacing 14 000 tonnes has KG, 11.0 m; KM, 12.0 m and is listed 3° to port. A tank of length, 10 m; breadth, 5 m; and 1 m deep with

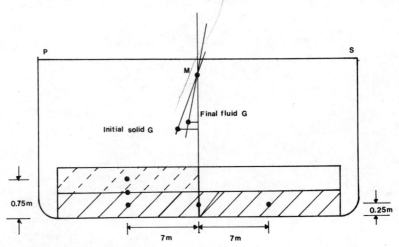

Figure 5.6 Example 5.6

centre of gravity, 7 m to port of the centre line is full of fresh water at Kg, 0.5 m. What will be the lift if half of this water is transferred to a similar tank on the starboard side of the vessel. The tanks are rectangular.

Free surface moment for one tank

$$= i\rho_{FW}$$

$$= \frac{1b^3}{12} \times 1 \text{ tonne m}$$

$$= \frac{10 \times 5^3}{12} \times 1 = 104.2 \text{ tonne m}$$

For two equal tanks

$$FSM = 208.4 \text{ tonne m}$$

$$\text{Weight of water transferred} = \text{volume} \times \rho$$
$$= 50 \text{ tonne}$$
$$G_0 G_1 = G_0 M \tan \theta$$
$$= 1 \tan 3^\circ$$
$$= 0.0524 \text{ m}$$

Weight	KG	Moment	G from centreline	Moment Port	Moment Starboard
14 000	11.0	154 000	0.052p	733.7	
− 25	0.75	− 19	7.000p	− 175.0	
25	0.25	6	7.000s		175.0
FSM		208			
14 000		154 195		558.7	175.0
					558.7
					383.7

$$KG_{\text{Fluid}} = \frac{\text{moment}}{\text{displacement}} \qquad G_0 G_1 = \frac{\text{moment}}{\text{displacement}}$$

$$= \frac{154 \, 195}{14 \, 000} \text{ m} \qquad = \frac{383.7}{14 \, 000} \text{ m} = 0.0274 \text{ m}$$

$$= 11.014 \text{ m}$$

$$KM = 12.000$$

$$\underline{GM_f = \quad 0.986 \text{ m}}$$

GM_{Fluid}

$$\tan \theta = \frac{G_0 G_1}{GM}$$

$$= \frac{0.0274}{0.982}$$

$$\theta = 1.6° \text{ port}$$

QUESTIONS ON FREE SURFACE

1. A vessel displacement 16 700 tonnes has KG, 9.4 m; KM, 10.0 m. She loads bunkers to a depth of 2.0 m in a rectangular tank of length, 10.0 m; breadth, 18.0 m; depth 3.0 m, subdivided by a single centre line division.
   RD of bunkers 0.90.
   Kg of bunkers 1.0 m.
   Find the fluid GM assuming KM remains constant.

2. A vessel has a fire in deck above a watertight flat. The area has length 20 m, breadth 18 m. The flat is 10 m above the keel.
   Given the following particulars, would it be safe to flood the flat to a depth of 3 m to extinguish the fire?
   Initial displacement, 11 500 tonnes.
   Initial KG, 9.7 m.
   KM at displacement, 12 607 tonne 10.8 m.

3. A vessel is displacing 10 000 tonnes and has solid KG, 9.0 m. There are four deep tanks athwartships. Each tank is 14 m long and 8 m wide.
   All the tanks contain palm oil RD 1.2. Find the fluid KG of the vessel.

4. A box-shaped vessel has length, 140 m; breadth, 20 m; draft, 8 m; KG, 7.5 m.
   What will be the fluid GM of the vessel if a midships well length 10 m extending the full breadth of the vessel is filled to a depth of 1 m by a breaking wave.
   The well deck is 10 m above the keel. Vessel in salt water.

5. A vessel displacing 20 000 tonnes has KG, 11 m and is floating in fresh water. She loads 500 tonnes of fresh water at Kg, 1 m into an

undivided double bottom tank length 30 m which has the following $\frac{1}{2}$ breadths at the surface of the liquid commencing from aft

Station	0	1	2	3	4	5	6
$\frac{1}{2}$ Ord.	3.3	7.0	8.3	9.4	10.4	11.1	11.6 m

Calculate the fluid *GM* of the vessel. *KM* at displacement 20 500 tonnes 12 m.

6. A vessel displacing 11 000 tonnes has *KG*, 8 m. A rectangular double bottom tank of length, 12 m; breadth, 6 m centre of gravity 4 m to starboard of the centreline is partially filled with 72 tonnes of fresh water *Kg*, 0.5 m. Find the list of the vessel. *KM* at displacement 11 072 tonnes 9.0 m.

7. A vessel displacement 18 500 tonnes has *KG*, 9.2 m; *KM*, 10.0 m and is listed 4° to port. She has on board a heavy lift of 100 tonnes *Kg*, 2 m stowed 6 m to starboard of the centreline.

    When the lift is being discharged it will be suspended from a derrick head 21 m above the keel 10 m to port of the centreline.

    Before commencing discharge a tank of length, 15 m; breadth, 9 m; depth, 4 m is filled to a depth of 2 m with fresh water.

    Kg of the water 1 m centre of gravity 4.5 m to starboard of the centreline.

    Find the list when the lift is suspended over side.

FREE SURFACE (ANSWERS)

1. Free of surface moment, 1093.5 tonne m. *KG* fluid, 9.304 m. *GM* fluid, 0.696 m.

2. Fluid *GM* after flooding 0.16 m.
    Very marginal and full residual stability calculation should be carried out.

3. Free surface moment, 2687.2 tonne m. *KG* fluid 9.287 m.

4. Free surface moment, 6833.3 tonne m. *KM* with well flooded 8.166 m. *GM* fluid, 0.344 m.

5. Free surface moment, 16 838.8 tonne m. Fluid *GM*, 0.423 m.

6. Free surface moment 216 tonne m. Fluid *GM*, 1.029 m. List, 1.43° port.

7. Free surface moment, 911.25 tonnes. *KG* fluid 9.231 m. List 5.66° port.

CALCULATION OF INITIAL *KG*—THE INCLINING TEST

In order to complete the assessment of initial stability the *KG* of the light ship must be found. Once light ship *KG* has been found any other *KG* can be determined by taking

$$KG = \frac{\text{light ship moment} + \sum \text{added weight moment}}{\text{displacement}}$$

$$\sum \text{added weight moment} = \text{deadweight moment}$$

$$\text{light ship moment} = \text{light displacement} \times \text{light } KG$$

Light displacement can be determined from hydrostatic data. However, it is impractical to determine light ship *KG* from first principles. Light ship *KG* is usually determined by an inclining test.

Inclining test

In *Figure 5.7 OR* is a pendulum of known length suspended above a horizontal graduated scale *PQ*. A small weight *w* is moved a distance of *m* across the deck of the vessel.

The centre of gravity of the vessel will move from G_0 to G_1 and the vessel will list until B_1 and G_1 are in the same vertical line. Provided the angle of heel is small, this vertical line will pass through *M*.

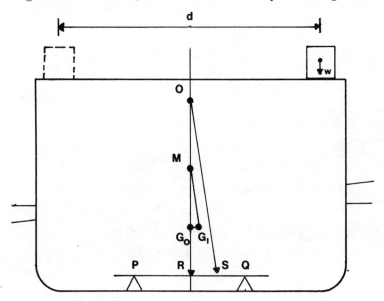

Figure 5.7 Inclining test

The pendulum will move until it has deflected to OS. Thus in triangles MG_0G_1, ORS

$$LO = LM = \text{angle of list}$$

$$LG_0 = LR = 90°$$

$\therefore$ Triangle MG_0G_1 is similar to triangle ORS

$$\therefore \quad \frac{G_0M}{G_0G_1} = \frac{RO}{RS}$$

$$G_0M = \frac{RO}{RS} \times G_0G_1$$

But if the vessel had displacement W

$$G_0G_1 = \frac{w \times d}{W}$$

$$G_0M = \frac{RO}{RS} \times \frac{w \times d}{W}$$

KM can be determined from ship form. Hence $KG_0 = KM - G_0M$.

Precautions to be taken during the test

1. *Conditions during test*
(a) The vessel should be as close as possible to the light condition.
(b) The vessel should be floating freely with moorings slackened during actual measurements.
(c) No other vessels moored alongside.
(d) Minimum wind and current.
(e) As few people as possible on board.
(f) All weights secured in seagoing position.
(g) No free surfaces.

2. *Conduct of test*
(a) Either more than one pendulum is used or they may be replaced by an accurate inclinometer.
(b) Several weights are moved in order so as to obtain the mean of several readings.
(c) The vessel must be upright at the commencement of the test.

3. *Calculation of results*
(a) Inclining weights used to bring the vessel upright to be accounted for.
(b) Weight of people on board to be taken into account.

(c) Any weights which will be part of the final displacement and not on board must be estimated and included in the light displacement.

(d) The displacement must be accurately determined.

A more detailed account of the calculation of inclining tests can be found in Reference 1.

Under operating conditions an inclining test may occasionally be used to give a quick check on the fluid *GM* of a loaded vessel. In this case a unit of cargo of known weight would be loaded on one side of the deck, the list noted, and the unit then moved to the other side and the change noted. An approximation to the fluid *GM* can then be calculated.

Example 5.7

An inclining experiment is carried out on a vessel completing fitting out. The following data is noted.

Displacement as inclined	9550 tonnes
Mass of inclining weights	10 tonnes
Distance weights shifted transversely	18 m
Length of plumb lines	9.5 m
Mean deflection of plumb lines	100 mm
Kg of inclining weights	12.5 m
KM of vessel from hydrostatic scales	8.35 m

Calculate the *GM* of the vessel as inclined. A tank containing 150 tonnes of fresh water is full at the time of the experiment. *Kg* of the water 1.0 m. Calculate the light ship *KG* of the vessel.

$$\frac{GM}{GG_1} = \frac{OP}{PQ}$$

$$GM = \frac{w \times d}{W} \times \frac{RO}{RS}$$

$$= \frac{10 \times 18}{9550} \times \frac{9.5}{0.1} \text{ m}$$

$$GM = 1.791 \text{ m}$$

$$KG = KM - GM$$

$$= 8.350 - 1.791 \text{ m}$$

$$= 6.559 \text{ m}$$

Weight	KG	Moment	Weight	KG	Moment
9550	6.559	62 642.2	10	12.5	125
160		275	150	1	150
9390		62 367	160		275

$$\text{Light ship } KG = \frac{62\,367}{9390} \text{ m}$$

$$= 6.642 \text{ m}$$

Example 5.8

While loading a cargo of timber on deck it is noted that a sling of timber weighing 8 tonnes KG, 12 m moved 16 m from one side of the ship to the other, inclines the vessel $1°$.

If the KM at this draft was 10.5 m calculate the approximate KG displacement 13 000 tonnes.

How much more deck cargo would it be safe to load if the GM was not to be less than 0.5 m.

$$GM = \frac{180 \times 8 \times 16}{1 \times \pi \times 13\,000}$$

$$KG = 10.5 - 0.564 = 9.936$$

Let x be cargo to be loaded at KG, 12 m.

Weight	KG	Moment
13 000	9.936	129 168
x	12.0	$12x$
13 000 + x		129 168 + 12x

$$10 = \frac{129\,168 + 12x}{13\,000 + x}$$

$$130\,000 + 10x = 129\,168 + 12x$$
$$2x = 832$$
$$x = 416 \text{ tonne}$$

UNRESISTED ROLLING IN STILL WATER

General treatment of ship motions is outside the scope of this book, however, one of the relationships which arises from the study of unresisted rolling in still water can be used to determine fluid GM.

If it is assumed that a vessel rolling through a small arc is executing simple harmonic motion with period T s (one period being from port to starboard and back to port) then if the radius of gyration of the vessel is K (see below) and g is the acceleration due to gravity.

$$T = \frac{2\pi \times K}{(gGM)^{\frac{1}{2}}}$$

From which

$$GM = \frac{4\pi^2 \times K^2}{g \times T^2}$$

The major difficulty in using this relationship is determining K. The parameter K arises from a general property of moments of inertia. A moment of inertia can always be expressed as

$$I = MK^2$$

where M is the mass of the body and K is the distance.

In the case of the equations above, the value of K is a radius of gyration which relates the moment of inertia of the vessel about a longitudinal axis through G, to the mass of the vessel.

For a light ship K can be determined at the same time as the inclining test. In theory K for any loaded condition can then be found by taking the sum of values of wd^2 where w is an added weight and d the distance from the centre of gravity G. In practice the process is very tedious. Sufficiently accurate values of K can be found by using empirical relationships. Reference 2 gives

$$\left(\frac{K}{B}\right)^2 = F\left(C_B C_U + 1.10 C_v 1 - C_B\right)\left(\frac{H}{d} - 2.20\right) + \frac{H^2}{B^2}\right)$$

where

$C_U =$ upper deck coefficient;
$H =$ effective depth of ship $= D + A/L_{pp}$;
$A =$ projected lateral area of erections;

L_{pp} = length between perpendiculars;
  d = mean draft;
  F = constant 0.125 for passenger and cargo ships;
    = 0.133 for tankers and bulk carriers.

For a typical ship this formula lends to values of K of about

$$K = 0.3B$$

Example 5.9

A vessel about to complete loading a timber deck cargo is induced to roll by landing a large sling of timber on the port side of the vessel.

If the period of the roll is 12.2 sec and K is 8 m find the GM of the vessel. As rolled the displacement of the vessel si 24 500 tonnes, and KM, 14 m.

How much more deck cargo can be loaded at Kg, 18 m if the KG of the vessel is to be no more than 12.50 m.

$$GM_F = \frac{4\pi^2 \times K^2}{g \times T^2} = \frac{4\pi^2 \times 8^2}{9.81 \times 12.2} = 1.73 \text{ m}$$

$$KM = 14.00 \text{ m}$$

$$\overline{KG = 12.27 \text{ m}}$$

Let w be deck cargo to load

Weight	KG	Moment
24 500	12.27	300 615
w	18.0	$18.0w$
24 500 + w		300 615 + 18.0w

$$KG = \frac{\text{moment}}{\text{weight}}$$

$$12.50 = \frac{300\,615 + 18.0w}{24\,500 + w}$$

$$12.50(24\,500 + w) = 300\,615 + 18.0w$$
$$306\,250 + 12.5w = 300\,615 + 18.0w$$
$$5.5w = 5635$$
$$w = 1025 \text{ tonnes}$$

Load 1025 tonnes at KG, 18.0 m.

QUESTIONS ON INCLINING TESTS

1. A vessel displacing 20 000 tonnes has a KM of 6.5 m. A weight of 25 tonnes shifted 20 m transversely across the deck causes a plumb line length 10 m to be deflected by 0.1 m.

    Kg of the inclining weight 10.5 m.

    If there remains to be installed a generator weighting 100 tonnes KG 3 m, determine the light ship KG.

2. A box-shaped barge of length, 100 m; breadth, 10 m; depth 6 m is floating at a draft of 2 m in fresh water.

    A weight of 1 tonne is moved 8 m transversely across the deck, deflecting a pendulum 5 m long by 0.05 m.

    What will be the GM of the vessel of loading a 200 tonne weight at a KG of 7 m?

3. When a weight of 20 tonnes is moved 12 m across the deck of the vessel to which the following data applies the vessel heels 0° 24'.

    Find the KG as inclined.

    Vessel floating at an even keel draft of 8 m in salt water. Underwater volume 15 000 m³, KB, 4.7 m. Length of 8 m waterplane 120 m. Half ordinates of the 8 m waterplane are

Station	0	1	2	3	4	5	6
½ Ord.	0	7	8	9	9	6	0 m

4. A vessel about to complete loading has yet to load 1500 tonnes of cargo at Kg, 8.0 m and Kg, 14.0 m. The displacement of the vessel is 14 000 tonnes and a partly filled double bottom tank containing fresh water has free surface moment 800 tonne m.

    A sling of cargo weighing 5 tonne already on the port side of the vessel is picked up and landed 16 m directly to starboard causing a change of list of 1.0°. A further 200 tonnes of fresh water is still to be loaded into the tank at Kg, 8.0 m in order to fill the tank.

    Distribute the cargo so that vessel has a departure GM of 6 m.

    KM at all relevant drafts 12.5 m.

    The vessel is floating in fresh water.

5. A vessel with a value of Kg of 11.4 m is observed to have a rolling period of 13.8 sec, when about to complete loading. KM as rolled 14.2 m, displacement 31 400 tonnes. A tank containing 1000 tonnes of ballast Kg, 0.5 m FSM 1250 tonnes m is to be emptied.

    How should a further 2400 tonnes of cargo be distributed between spaces Kg, 5 m and Kg, 19 m so that the final KG is to be no more than 11.5 m.

INCLINING TEST (ANSWERS)

1. KG as inclined 4 m
   Light ship, KG 3.987 m

2. KG as inclined 4.76 m
   GM after loading weight -0.08 m (unstable)

3. $I_{CL} = 33\,928.9\ m^4$
   $KM = 6.962$ m
   $KG = 4.726$ m

4. KG as inclined 12.17 m
   KG_{solid} 12.11 m
   Load 855 tonne at Kg 14 m
        645 tonne at Kg 8 m

4. GM as rolled 2.75 m
   KG as rolled 11.45 m
   Load 1870 tonnes at Kg 5 m
        530 tonnes at Kg 19 m

INCREASE IN DRAFT DUE TO HEEL (*Figure 5.8*)

A vessel draft d is heeled or listed to some angle θ. At angle θ the maximum draft of the vessel is $a + b$.

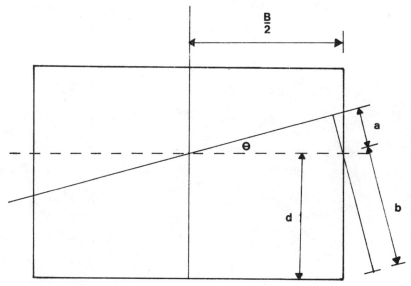

Figure 5.8 Increase in draft due to heel

Then if the vessel has breadth B

$$a = \frac{B}{2} \sin \theta$$

$$b = d \cos \theta$$

$\therefore$ Draft heeled $= \dfrac{B}{2} \sin \theta + d \cos \theta$

This relationship can be adapted to take into account rise of floor and turn of bilge. However, the indication of increased draft given by the equation will be sufficient for most practical purposes.

References

1. RAWSON and TUPPER, *Basic Ship Theory*, Chapter 4
2. KATO, C, *On the approximate calculation of a ship's rolling period, Japanese Society of Naval Architects Journal*, **89** (1956)

6 Stability at large angles of heel

This chapter deals with the assessment of stability of large angles of heel, when the assumptions about the transverse metacentre used in Chapters 4 and 5 are no longer valid. Regulations about merchant vessel stability are introduced.

OBJECTIVES

1. To develop the form of the GZ curve as the vessel is heeled to any angle of heel.
2. To present the conditions of equilibrium as GZ curves.
3. To assess dynamical stability.
4. To assess the stability condition of a vessel with respect to Load Line Rules and Grain Rules.
5. To describe the effect of changes of form on the stability of a vessel.
6. To assess the effect of wind on the stability of container vessel.

FORM OF GZ CURVES

When conditions of equilibrium were discussed, it was shown that the stability condition of the vessel depended upon the sense of the couple formed by displacement acting through the centre of gravity and buoyancy acting through the centre of buoyancy.

Since the magnitude of the force of displacement is constant at all angles of heel, the size of the couple depends upon the arm of the couple, measured from the centre of gravity G to a point Z on the vector through B. Therefore the measurement of the value of GZ at all angles of heel gives an indication of the stability of the vessel.

For a ship shape the values of GZ will have the following pattern for a stable vessel.

1. *Figure 6.1(a)*. Suppose the vessel is heeled to some angle θ_1 then the centre of buoyancy will move from B_0 to B_θ as the result of shift of

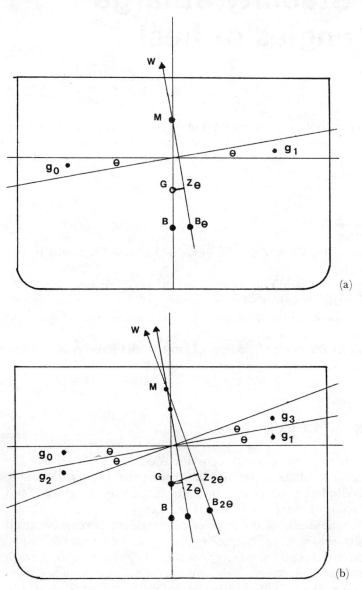

Figure 6.1 *Increase in GZ up to angle of deck edge immersion*

buoyancy from g_0 to g_1. Now if the vessel is heeled through a further angle θ_2 $(\theta_1 = \theta_2)$ (*Figure 6.1(b)*), then provided the deck edge is not immersed, or the bilge emersed, it can be seen that a larger volume of buoyancy will move from g_2 to g_3 and that $g_2g_3 > g_0g_1$. Hence the shift of the centre of buoyancy $BB_\theta > B_\theta B_{2\theta}$. Hence the value of GZ will be increasing and increasing more rapidly as the vessel is heeled.

Note that if $BB_\theta > B_\theta B_{2\theta}$ then the metacentre M can no longer be a fixed point on the centreline.

The value of GZ will continue to increase until either the deck edge is immersed or the bilge is emersed.

2. *Figure 6.2.* When the deck edge is immersed the condition will change. It is no longer possible to assume that the buoyancy is transferred equally from one side of the centreline to the other. The condition that underwater volume is constant must be maintained. Hence the emersed volume of buoyancy must equal the immersed volume of buoyancy.

Note if the waterlines are drawn so as to intersect at the centreline, then a volume represented by $a.b.c$ must be replaced by the vessel 'sinking' until a layer of buoyancy $a.d.e.f.$ is immersed. Clearly the centroid of this layer is on the emersed side of the centreline. Hence the shift of buoyancy is reduced. The rate of increase of GZ will be reduced.

After deck edge immersion it becomes difficult to describe the

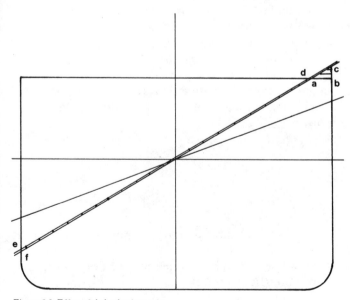

Figure 6.2 Effect of deck edge immersion

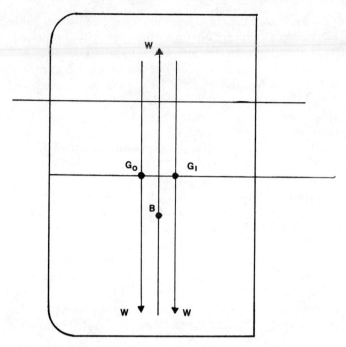

Figure 6.3 Stability at 90°

behaviour of GZ, particularly as the bilge emerges. However, if we consider the extreme case of the vessel at 90° heel then the sense of the theoretical couple depends upon the relative positions of G and B (assuming G has not moved), i.e. if in *Figure 6.3* the centre of gravity is at G_0 the vessel is stable at 90° while if it is at G_1 the vessel is unstable.

If this pattern of movement of B relative to G is plotted as values of GZ against angle of heel a typical curve of statical stability (GZ curve) is obtained. (*Figure 6.4, 6.5(a), (b), (c), (d)*).

FEATURES OF GZ CURVES

Initial slope and GM

A knowledge of the initial GM can be used to determine the slope of the origin of the GZ curve.

In *Figure 6.6*, AD is a line drawn as a tangent to the origin of the GZ curve. AD cuts an ordinate DE erected at 57.3° heel. BC is drawn close to the origin at angle θ.

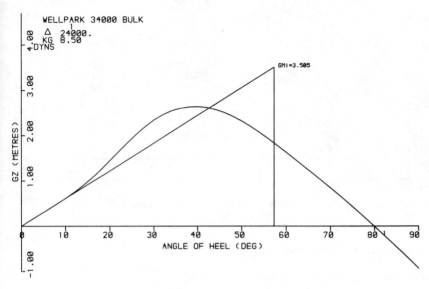

Figure 6.4 GZ curve of a typical stable vessel

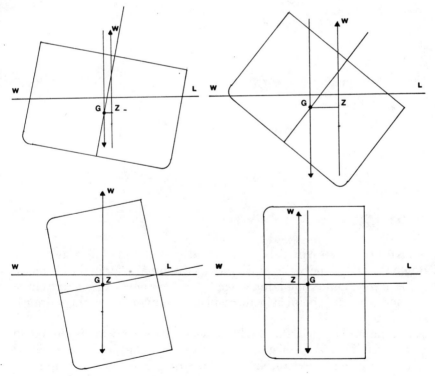

Figure 6.5 Change in GZ with heel

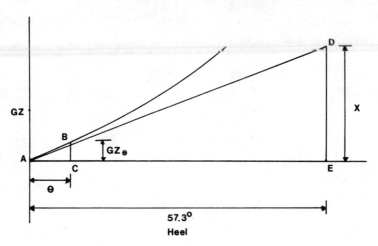

Figure 6.6 Relationship between GM and GZ curve

Triangle ABC is similar to triangle ADE

$$\frac{DE}{AE} = \frac{BC}{AC}$$

$$\frac{DE}{57.3} = \frac{GZ\theta}{\theta}$$

Now if θ is small

$$GZ = GM \sin \theta$$

and $\sin \theta \simeq \theta_c$ (θ measured in radians)

and $57.3° = 1$ radian

$$\therefore \quad \frac{DE}{1} = \frac{GM\theta_c}{\theta_c}$$

$$DE = GM$$

Thus if an ordinate equal to GM is erected at $57.3°$ and a line drawn to the origin then the slope of this line indicates the initial slope of the GZ curve.

Deck edge immersion is indicated by the point of flexure of the curve, although the exact point of flexure will depend upon sheer and position of superstructure.

Maximum GZ and the angle at which it occurs can be found by inspection. The range of stability can also be found by inspection.

It must be emphasized that only the early part of the curve up to say $40°$ heel can be regarded as giving a reasonable representation of the actual GZ value, as in practice at very large angles of heel it is probable that:

(a) Cargo will have shifted.
(b) Equipment will have broken loose.
(c) Water will have entered the vessel.

Thus making invalid the assumption that G does not shift.

DYNAMICAL STABILITY

Before dealing with regulations about stability it is necessary to be able to measure dynamical stability using the GZ curve.

In the general discussion on conditions of equilibrium, the concept of dynamical stability was presented on the rise in the centre of gravity of a body as it was disturbed. In the case of a floating body the concept can be considered in terms of the separation between the centre of gravity G and the centre of buoyancy $B_0 B_1$—B_θ as a vessel which is initially stable is heeled.

In *Figure 6.7* a vessel which is initially stable is heeled through a small angle θ from the upright. If the initial separation between G_0 and B_0 is d_0.

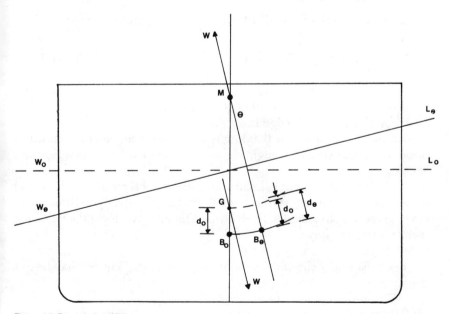

Figure 6.7 Dynamical stability

Then if we assume that the metacentre M remains fixed on the centreline and centre of buoyancy will move in an arc centre M radius MB_0. If the vertical separation between G_0 and B_0 is now measured and is d_0, it can be seen that $d_0 > d_0$. Hence work has been done separating G and B.

Dynamical stability $= W(d_0 - d_0)$

The problem is to determine this value in general. If we consider a vessel heeled to some angle θ and then heeled through a further small angle $\delta\theta$. Then in *Figure 6.8(a)* and *(b)* the increase in separation of G and B_0 is $GZ_\theta\delta\theta$

$$\therefore \quad (d_0 - d_1) = \sum_{GZ=0}^{GZ=0} GZ\,\delta\theta$$

If on GZ curve (*Figure 6.9*) $GZ_0\delta\theta$ is plotted it can be seen that $GZ\delta\theta$ is an element of the GZ curve. Therefore in the limit

$$(d_0 - d_1) = \int_0^H GZ\,d\theta$$

which is the area beneath the GZ curve up to angle Φ

$$\therefore \quad \text{Dynamical stability} = W \text{ (Area beneath } GZ \text{ curve)}$$

STABILITY REQUIREMENTS UNDER LOAD LINE RULES

These regulations, as well as others, define the stability of a vessel by defining the GZ curve up to either 40° or the angle of progressive flooding and sometimes to deck edge immersion.

The angle of progressive flooding is taken as the angle at which water could enter the vessel through say an airpipe or other non-watertight opening.

The Merchant Shipping (Load Line) Rules 1968 require that a vessel should have satisfactory stability in all probable loading conditions, and when loaded to the assigned freeboards. The rules lay down the following specific criteria (*Figure 6.10*):

(a) The area under the curve of Righting Levers (GZ curve) shall be not less than:
(i) 0.055 , radians up to an angle of 30°.
(ii) 0.09 m radians up to an angle of either 40° or the angle at which the

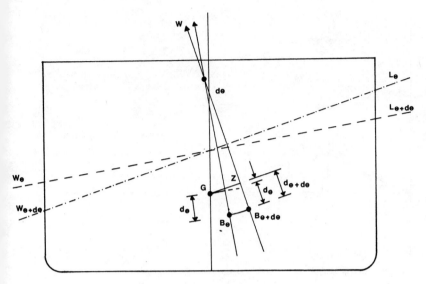

Figure 6.8(a) Relationship between dynamical stability and GZ

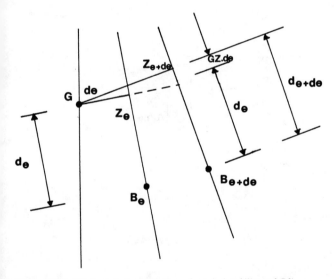

Figure 6.8(b) Detail of relationship between dynamical stability and GZ

lower edges of any openings in the hull superstructure of deck house being openings which cannot be closed weathertight are immersed if that angle be less.

(iii) 0.03 m radians between the angles of 30° and 40° or such lesser angle referred to in (ii).

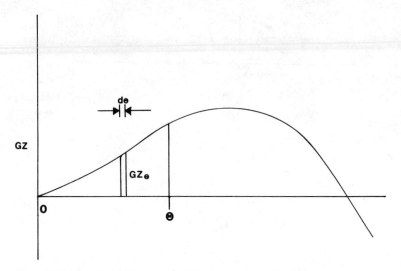

Figure 6.9 Relationship between area under GZ curve and dynamical stability

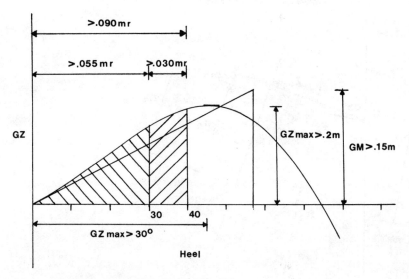

Figure 6.10 Stability requirements of 1968 Load Line Rules

(b) The Righting Lever (*GZ*) shall be at least 0.20 m at an angle equal to or greater than 30°.

(c) The maximum Righting Lever (*GZ*) shall occur at an angle of heel of not less than 30°.

(d) The initial transverse metacentric height shall be not less than 0.15 m. In the case of a ship carrying a timber deck cargo which complies with (a)

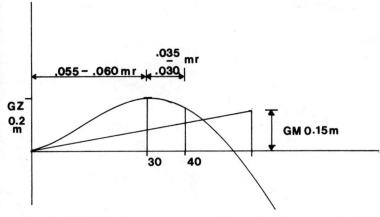

Figure 6.11 Minimum curve

by taking into account the volume of timber deck cargo the initial transverse metacentric height shall not be less than 0.05 m.

The rules also require the vessel to undergo an inclining test to determine whether or not the vessel complies with the requirements (Ref. SI 1968 No. 1053 Schedule 4 Part 1 2.(2)(a)–(d)). The minimum curve would have the form shown in *Figure 6.11*, although as drawn the area under the curve between 30° and 40° would probably be deficient.

CONDITIONS OF EQUILIBRIUM FROM GZ CURVES

The curves shown so far all indicate a vessel in initial stable equilibrium. For vessels in other conditions of equilibrium the curves would appear as follows:

Unstable initial equilibrium (*Figure 6.12*)

The vessel may go to an angle of loll at which point the vessel is stable relative to the angle of loll. Note that the initial negative slope is very shallow and the $\div GM$ small. The maximum GZ is small and the range of positive stability small. The vessel is in a dangerous condition.

This condition can be corrected by lowering the centre of gravity. There are three possible methods

(a) Lowering weights, generally impractical as loll is much more likely in a full vessel rather than an empty one.

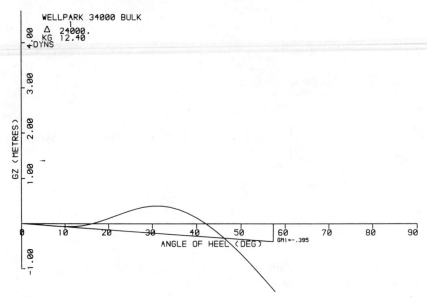

Figure 6.12 GZ curve of vessel in unstable equilibrium

(b) (i) Filling a subdivided double bottom tank on the low side of the vessel. This has the effect of keeping the vessel over to one side while G is lowered and keeps free surface to a minimum. (ii) Bringing the vessel upright by filling a tank on the high side.

This method reduces freeboard and if the vessel has only a small reserve of stability will cause a large angle of list with consequent risk of cargo shift.

(c) Removing weight from the high side. This in practice means jettisoning cargo from the deck or removing ice. The effect is the reverse of (b) except that freeboard is increased which is an advantage.

The most important lesson to learn from these procedures is that it is much better not to loll in the first place.

Neutral initial equilibrium *(Figure 6.13)*

The curve is initially horizontal, however, it should be noted that in most cases the curve will slope upwards after a few degrees of heel and the vessel would have some range of positive stability.

In *Figures 6.14(a)*, *(b)* and *(c)* the curves are redrawn to show the effect of heeling to port and starboard.

The following properties can be noted:

1. The vessel is stable whenever the GZ curve is in a positive quadrant.

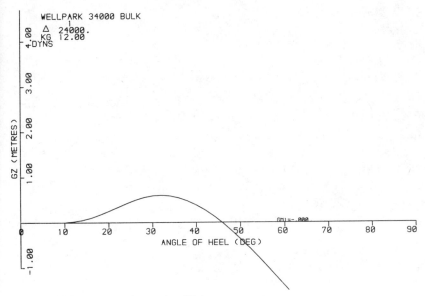

Figure 6.13 GZ curve of vessel in neutral equilibrium

2. The vessel moves into stable equilibrium whenever the curve crosses the heel axis in a positive direction.
3. The vessel is in unstable equilibrium when the curve crosses the line in a negative direction.

EFFECT OF FORM ON *GZ* CURVE

The *GZ* curves used to illustrate this section are based on computer outputs for the *GZ* curves of box shapes.

1. *Change in freeboard (Figures 6.15 and 6.16)*
Suppose a vessel has centre of gravity at G_0 and freeboard f_0, can have additional freeboard added to give freeboard f_1 with the draft remaining constant and the centre of gravity remaining at G_0. (*G* can remain fixed if weight is redistributed.)

Then if the vessel is heeled by an external force the initial shape of the curve will be unchanged. However, the angle of deck edge immersion will be delayed for the vessel with high freeboard. Thus the curve will continue to rise, until the larger angle of deck edge immersion. There is a considerable increase in max. *GZ*, the range of stability is increased and at large angles of heel the dynamical stability in increased. The improved stability at very large angles can be accounted for by considering the increased width of waterplane.

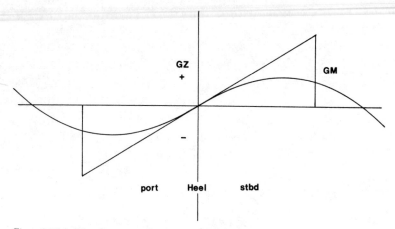

Figure 6.14(a) GZ curve to port and starboard of upright for stable vessel

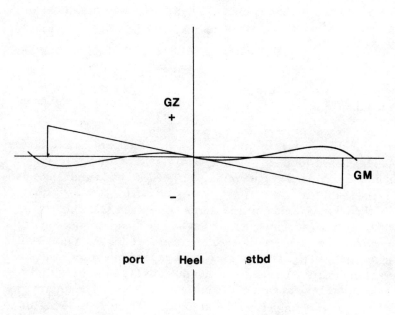

Figure 6.14(b) GZ curve to port and starboard for vessel in unstable equilibrium

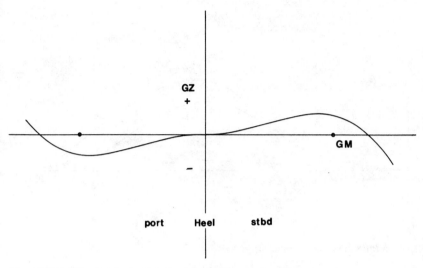

Figure 6.14(c) GZ curve to port and starboard for vessel in neutral equilibrium

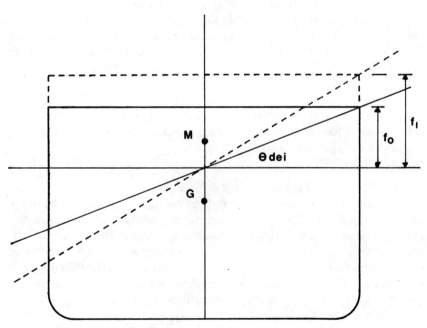

Figure 6.15 Effect of stability of increasing freeboard

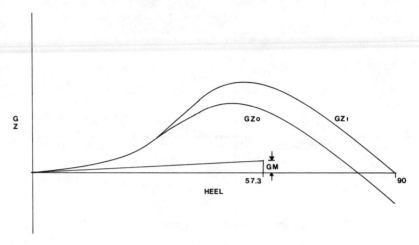

Figure 6.16 Effect on GZ curve of increasing freeboard

2. Superstructure (*Figure 6.17*)

Watertight superstructure has a similar effect to increased freeboard, provided the superstructure is distributed equally about the centre of buoyancy, i.e. forecastle and poop or uniformly stowed timber deck cargo, etc.

However, if the superstructure is not uniform, i.e. offshore supply vessels, there will be a considerable shift of the centre of buoyancy when the superstructure enters the water causing the vessel to trim. As in the figure the trim will be towards the part of the vessel with the lower freeboard. This change of trim due to shift buoyancy as a vessel is heeled is called free trim.

All vessels will trim to some extent as they are heeled, in most cases the effect is ignored in presenting GZ curves. The use of computers makes the calculation of free trim practical and some programmes for calculating GZ take free trim into account.

3. Increase in beam (*Figures 6.18 and 6.19*)

Suppose a vessel with centre of gravity at G_0 and beam B_0 has its beam increased to B_1. In this case the centre of gravity can remain at G_0. However displacement must be increased. Since the width of the waterplane is increased, the inertia of the waterplane must be increased, hence the metacentre will rise to M_1. The initial slope of the GZ curve will be increased. However the angles of deck edge immersion will be earlier, and thereafter the slope of the curve will be reduced.

At large angles the waterplane is not greatly changed, there is little change in stability at these angles, the curves coincide at some very large

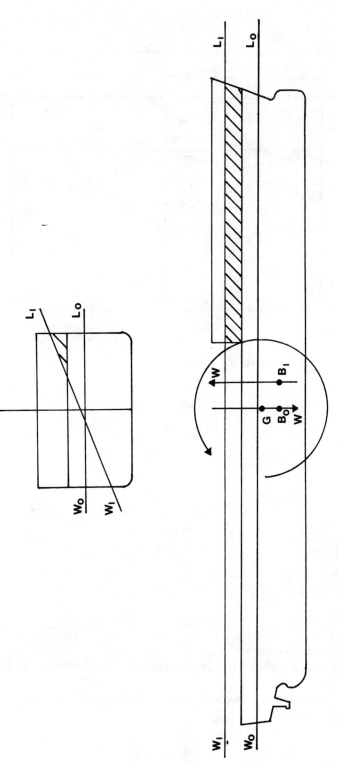

Figure 6.17 Effect of superstructure on stability

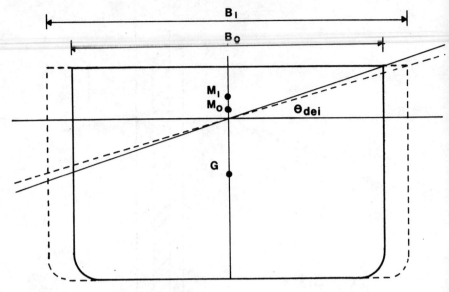

Figure 6.18 *Effect on stability of increasing beam*

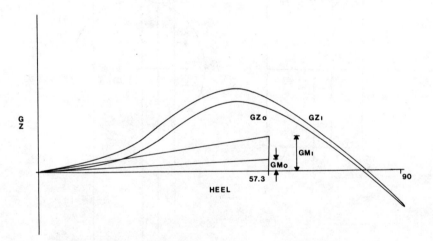

Figure 6.19 *Effect on GZ curve of increasing beam*

angle. In the special case of a box shape the curves intersect at 90°; for most other vessels the point of intersection will be at some angle less than 90°.

Since displacement has been increased there will be an increase in righting moment.

Table 6.1

	Increased freeboard	Even super-structure	Uneven super-structure	Increased beam	Increased length
Displacement	No change	No change	No change	Increase	Increase
GM	No change	No change	No change	Increase	No change
Deck edge immersion	Increased	Effective increase	Effective increase	Reduced	No change
Max. GZ	Increased	Increased	Increased	Small increase	No change
Range of stability	Increased	Increased	Increased	Possibly increased or decreased	No change
Righting moment	Increase at large angles	Increase at large angles	Increase at large angles	Increase	Increase
Dynamical stability to 40°	Increase	Increase	Increase	Increase	Increase
Trim	No change	Small change	Large change	No change	No change
Angle of flooding	Increased	Possible small reduction	Possible large reduction	Possible reduction	No change

4. *Length*

Length can be increased without altering the position of G and there will be no effect on the value of KM for box shapes and little effect for ship shapes. There will be no change in the angle of deck edge immersed, the shape of the GZ curve will be little changed. However, displacement must be increased thus increasing both righting moment and dynamical stability. The effects of changed form can be summarized in *Table 6.1*.

EFFECT OF SHIFTING WEIGHTS WITHIN THE VESSEL

Vertical shift of weight

In *Figure 6.20* a vertical shift of weight has caused the centre of gravity to rise from G_0 to G_1. If the vessel is heeled to some angle θ then the value of the righting lever has been reduced from G_0Z_0 to G_1Z_1.

$$G_1Z_1 = G_0Z_0 - G_0X$$
$$G_0X = G_0G_1 \sin \theta$$
$$G_1Z_1 = G_0Z_0 - G_0G_1 \sin \theta$$

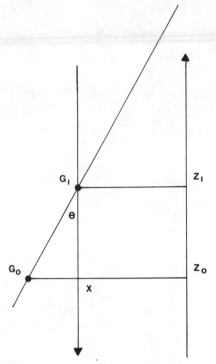

Figure 6.20 Effect on GZ of moving G vertically

and if the centre of gravity is lowered

$$G_1 Z_1 = G_0 Z_0 + G_0 G_1 \sin \theta$$

The effect on stability is shown in *Figure 6.21*.

The rise in G therefore reduces the range of stability, the maximum GZ and reduces GM by $G_0 G_1$.

In *Figure 6.22* the effect of $G_0 G_1 > G_0 M$ is shown. During the early part of the curve

$$G_0 G_1 \sin \theta > G_0 Z$$

hence the vessel is lolling. The vessel reaches the angle of loll when

$$G_0 G_1 \sin \theta = G_0 Z_0$$

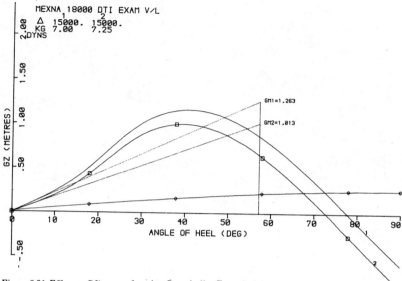

Figure 6.21 Effect on GZ curve of moving G vertically. Example 6.1

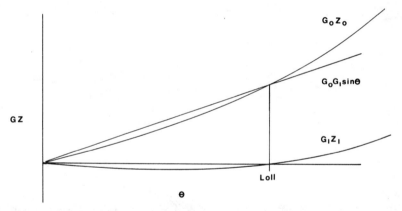

Figure 6.22 Shift of weight vertically causing instability

Horizontal shift of weight

In *Figure 6.23* a horizontal shift of weight has caused the centre of gravity to move from G_0 to G_1. If the vessel is heeled to some angle θ then the value of the righting lever has been reduced from G_0Z_0 to G_1Z_1.

$$G_1Z_1 = G_0Z_0 - G_0X$$

$$G_0X = G_0G_1 \cos \theta$$

$$G_1Z_1 = G_0Z_0 - G_0G_1 \cos \theta$$

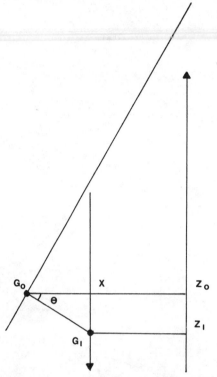

Figure 6.23 Effect on GZ of moving weight horizontally

The effect on stability is shown in *Figure 6.24*. The value of GZ is reduced at all angles of heel.

The vessel moves to an angle of list where

$$G_0 G_1 \cos \theta = G_0 Z_0$$

The range of stability on the low side will be reduced and dynamical stability will be reduced. The vertical separation between G and M when the vessel is upright is unchanged.

Example 6.1 (*Figure 6.21*)

A vessel displacing 15 000 tonnes has KG, 7 m. Cargo is redistributed to cause KG to rise by 0.25 m. The values of GZ in the initial condition where

Heel	0	15	30	45	60	75	90	degrees
GZ	0.00	0.391	1.000	1.138	0.774	0.129	−0.584	m

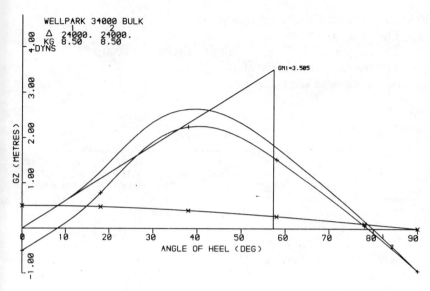

Figure 6.24 Effect on GZ curve of moving weight horizontally

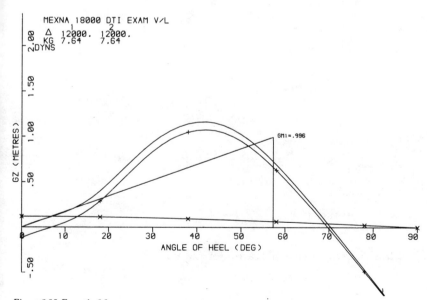

Figure 6.25 Example 6.2

Find for the initial and final condition
(a) The range of stability;
(b) The maximum GZ and the angle at which it occurs;
(c) The dynamical stability at $40°$;
(d) An estimate of the GM.

Heel	0	15	30	45	60	75	90
$G_0 Z_0$	0.000	0.391	1.000	1.138	0.774	0.129	−0.584 m
$G_0 G_1 \sin$ heel	0.000	−0.064	−0.125	−0.177	−0.217	−0.242	−0.250 m
$G_0 Z_1$	0.000	0.327	0.875	0.961	0.557	−0.113	−0.834 m

$G_0 M_0$ 1.26 m   Range $78°$
$G_1 M_0$ 1.01 m   Range $72°$
Max. $G_0 Z_0$ 1.20 m at $42°$
Max. $G_1 Z_1$ 1.00 m at $40°$

Dynamical stability (initial)

Heel	GZ	SM	F (Area)
0	0.00	1	0
10	0.24	4	0.96
20	0.60	2	1.20
30	1.00	4	4.00
40	1.15	1	1.15
			———
			7.31

$$\text{Area} = \frac{1}{3} \times h \times \sum F \text{ (Area)}$$

$$= \frac{1}{3} \times \frac{10}{57.3} \times 7.31 \text{ m radians}$$

$$= 0.43 \text{ in radians}$$

Dynamical Stability $= W \times$ (Area under GZ curve)

$$= 15\,000 \times 0.43 \text{ tonne m}$$

$$= 6450 \text{ tonne m}$$

Dynamical stability (final)

Heel	GZ	SM	F (Area)
0	0.00	1	0.00
10	0.16	4	0.64
20	0.50	2	1.00
30	0.88	4	3.52
40	1.00	1	1.00
			6.16

$$\text{Area} = \frac{1}{3} \times h \times \sum F \text{ (Area)}$$

$$= \frac{1}{3} \times \frac{10}{57.3} \times 6.16 \text{ m radians}$$

$$= 0.36 \text{ m radians}$$

$$\text{Dynamical stability} = W \times (\text{Area under } GZ \text{ curve})$$

$$= 15\,000 \times 0.36 \text{ tonne m}$$

$$= \frac{5700}{5400} \text{ tonne m.}$$

Example 6.2 *(Figure 6.25)*

A vessel displacing 12 000 tonnes has KG, 7.64 m. The values of GZ are as follows:

Heel	0	10	20	30	40	50	60	70	degrees
GZ	0	0.19	0.50	0.94	1.16	1.03	0.60	0.06	m

Cargo is redistributed so as to make the centre of gravity 0.13 m to port. Find the list and the dynamical stability from the list angle to 30°.

Heel	0	10	20	30	40	50	60	70
$G_0 Z_0$	0	0.19	0.50	0.94	1.16	1.03	0.60	0.06
$G_0 G_1 \cos$	0.13	0.13	0.12	0.11	0.10	0.08	0.07	0.05
$G_1 Z_1$	−0.13	0.06	0.38	0.83	1.06	0.95	0.53	0.01

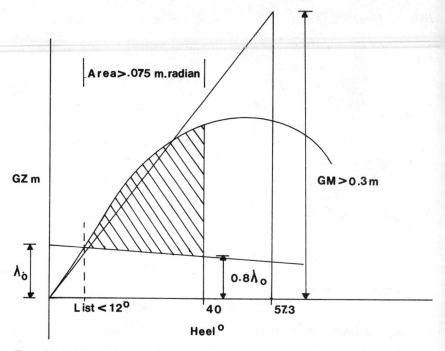

Figure 6.26 Stability requirements under Grain Rules

From *Figure 6.26*

Heel 7°

Area 10° – 30°

Heel	GZ (m)	SM	F (Area)
10	0.06	1	0.06
20	0.38	4	1.52
30	0.83	1	0.83
			2.41

$$\text{Area } 10\text{--}30 = \frac{1}{3} h \times \sum F \text{ (Area)}$$

$$= \frac{1}{3} \times \frac{10}{57.3} \times 2.41 \text{ m radians}$$

$$= 0.140 \text{ m radians}$$

Area 7–10 assuming that area is triangular

$$\text{Area} = \frac{1}{3} \times \frac{3}{57.3} \times 0.06 \text{ m radians}$$

$$= 0.001 \text{ m radians}$$

Area 7–30 = 0.141 m radians

$$\begin{aligned}
\textit{Dynamical stability} &= \text{Displacement} \times \text{area} \\
&= 120\,000 \text{ tonnes} \times 0.141 \text{ m radians} \\
&= 1692 \text{ tonne m.}
\end{aligned}$$

GRAIN

When a vessel is loaded with grain, the calculation of her stability conditions has to take into account the possibility of horizontal and vertical shift of the grain. The loading of grain is subject to regulations.

The grain regulations define the intact stability requirements for any ship carrying grain in bulk as follows[2] (*Figure 6.26*).

Intact stability requirements

In all ships to which these Regulations apply, the intact stability characteristics of any ship carrying grain in bulk shall be shown to meet throughout the voyage at least the following criteria:

(i) the angle of heel due to the shift of grain shall be not greater than 12° unless, in relation to a particular ship or class of ship a lesser angle is required by an Administration;
(ii) in the statical stability diagram, the residual area between the heeling arms curve and the righting arm curve up to the angle of maximum difference between the ordinates of the two curves or 40° or the angle of flooding (θ_r) whichever is the least shall in all conditions of loading be not less than 0.075 m radians and
(iii) the initial metacentric height after correction for the free surface of liquids in the tanks shall be not less than 0.30 m.

The angle of flooding (θ_r) is the angle of heel at which openings in hull, superstructures or deck houses, which cannot be closed weather tight immerse. Small opening through which progressive flooding cannot take place can be disregarded.

The vessel is to be upright on completion of loading.

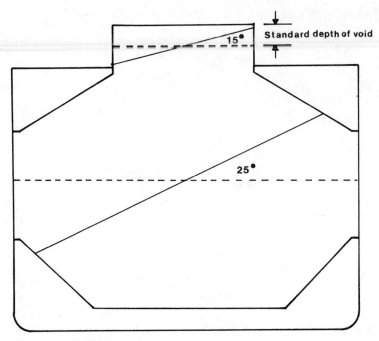

Figure 6.27 Assumed shift of grain

The heeling arm curve mentioned in (ii) above is calculated from information about the assumed shift of grain within the vessel. The detailed assumptions are given as Reference 2. The regulations assume that for 'filled' compartments there will be a void space beneath the top of the compartment, the depth of the void being determined by formula which depends the distance from the hatch to the compartment boundary, and the depth of the hatch side girder.

The grain in the filled compartment is then assumed to shift 15° to the horizontal (*Figure 6.27*). The Volumetric Heeling Moment (VHM) can then be calculated from the geometry of the compartments. In most cases the volumetric centre of the compartment will be taken as the *KG* of the grain. However, the *KG* is lowered by taking into account the void spaces then the vertical shift of grain must be taken into account and then

total heeling moment = 1.06 × calculated heeling moment

In the case of partly filled compartments, *Figure 6.27* the grain is assumed to shift 25° to the horizontal before the VHM is calculated, when

total heeling moment = 1.12 calculated heeling moment

The sum of all the values of VHM is taken for a particular load condition and the value λ_0 calculated from:

$$\lambda_0 = \frac{\text{Assumed volumetric heeling moment}}{\text{Stowage factor} \times \text{displacement}}$$

λ_0 is equivalent to

$$G_0G_1 = \frac{\text{Weight of grain shifted} \times \text{distance shifted}}{\text{Displacement}}$$

The value $\lambda_{40} = 0.8\lambda_0$

The values of λ_0 plotted at $0°$ and λ_{40} plotted at $40°$ give the two points through which the heeling arm curve is drawn. Comparing this process with plotting the general heeling arm curve $G_0G_1 \cos \theta$

$$G_0G_1 = \lambda_0$$
$$G_0G_1 \cos 40° = G_0G_1 \times 0.766$$
$$\simeq 0.8\lambda_0$$

Hence the assessment of stability under the grain rules is a rather complex application of finding list due to transverse shift of weight, noting that VHM is very similar to free surface except that it is calculated for two particular angles of heel.

Example 6.3 (*Figure 6.28*)

A vessel displacing 16 500 tonnes has KG, 7.50 m. Calculations of grain shift give the following data.

$$\text{Volumetric heeling moment} = 3960 \text{ m}^4$$
$$\text{Stowage factor} = 1.2 \text{ m}^3/\text{tonne}$$

For the given displacement and KG the vessel has the following values of GZ at the angles of heel given.

Heel	0	15	30	45	60	75	90
GZ	0.00	0.267	0.645	0.571	0.163	−0.454	−1.104 m

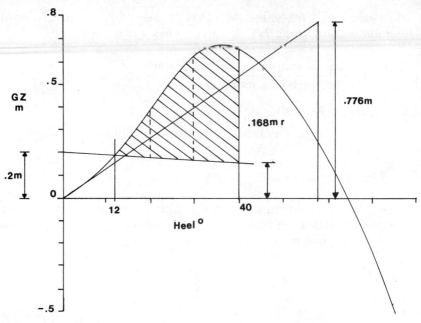

Figure 6.28 Example 6.3

$$\lambda_0 = \frac{\text{Volumetric Heeling Moment}}{\text{Displacement} \times \text{S.F.}}$$

$$= \frac{3960}{16\,500 \times 1.2}\ \text{m}$$

$$= 0.20\ \text{m}$$

$$\lambda_{40} = 0.8\lambda_0$$

$$= 1.16\ \text{m}$$

From curve

Area $20° - 40°$

Heel	GZ (res)	SM	F (Area)
20	0.21	1	0.21
30	0.48	4	1.92
40	0.50	1	0.50
			2.63

$$\text{Area} = \frac{1}{3} \times \frac{10}{57.3} \times 2.63 \text{ m radians}$$

$$= 0.153 \text{ m radians}$$

Area $12° - 20°$

This area can be regarded as being approximately triangular

$$\text{Area} = \frac{1}{2} \times \frac{8}{57.3} \times 0.21 \text{ m radians}$$

$$= 0.015 \text{ in radians}$$

$$\begin{aligned}\text{Area } 12 - 40 &= (0.153 + 0.015) \text{ m radians} \\ &= 0.168 \text{ m radians}\end{aligned}$$

Initial GM $= 0.776$ m satisfactory

Heel $= 12°$ marginal

Residual Area $= 0.168$ m radians satisfactory

Example 6.4 (*Figures 6.29 and 6.30*)

The *MV Nonesuch* has light displacement 7304 tonnes and *KG*, 10.09 m. The data in *Figure 6.29* can be assumed to apply to all compartments. She loads grain SF 1.3 m³/tonnes as follows:

> No 1 Full
> No 2 Ullage 8 m
> No 3 Full
> No 4 Full
> No 5 Ullage 2 m
> No 3 DB Oil Fuel 1360 tonnes *KG*, 1.67 m
> Fresh Water Tank 284 tonnes *KG*, 12.72 m

Does the vessel comply with the Grain Regulations?

Comp	Volume	SF	Weight	KG	Moment	VHM
Lt Ship			7304	10.09	73 697	
1	7860	1.3	6046	7.90	47 763	823
2	4050	1.3	3115	4.9	15 264	14 600
3	7860	1.3	6046	7.90	47 763	823
4	7860	1.3	6046	7.90	47 763	823
5	7300	1.3	5615	7.40	41 551	2100
No 1 3 DB			1360	1.67	2271	
FW			284	12.12	3612	
			35 816	12.72	279 684	19 169

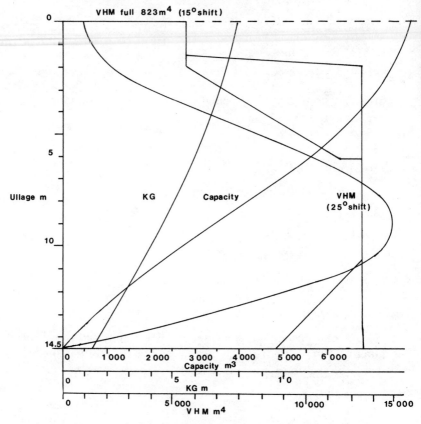

Figure 6.29 Grain and stability data for holds of MV Noneusch

$$KG = \frac{\text{Moment}}{\text{Weight}} = \frac{279\,684}{35\,816} = 7.809 \text{ m}$$

$$KM \qquad\qquad = 11.300 \text{ m}$$

$$GM \qquad\qquad = 3.491 \text{ m}$$

Heel	KN	KG sin	GZ
0	0	0	0
5	1.00	0.68	0.38
12	2.42	1.62	0.80
15	2.90	2.02	0.88
30	5.60	3.90	1.70
45	7.30	5.52	1.78
60	8.05	6.76	1.29
75	8.20	7.54	0.66
90	7.60	7.81	−0.21

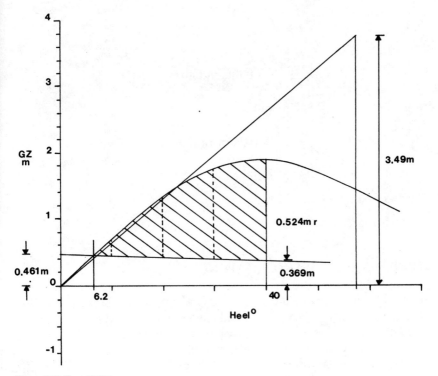

Figure 6.30 Example 6.4

$$\lambda_0 = \frac{VHM}{\text{displacement} \times SF}$$

$$= \frac{2469 \times 1.06 + 16\,700 \times 1.12}{35\,816 \times 1.3}$$

$$= 0.46 \text{ m}$$

$$0.8\lambda_0 = 0.37 \text{ m}$$

GM	3.491 m	Min. *GM*	0.300 m
List	6.2°	Max. List	12°

Area

Station	Ord.	SM	F (Area)
10	0.24	1	0.24
20	0.91	3	2.73
30	1.20	3	3.60
40	1.32	1	1.32
			7.89

$$\text{Area } 10-40 = \frac{3}{8} \times h \times \sum F \text{ (Area)}$$

$$= \frac{3}{8} \times \frac{10}{57.3} \times 7.89 \text{ m radians}$$

$$= 0.516 \text{ m radians}$$

$$\text{Area } 6.2-10 = \frac{3.8}{57.3} \times \frac{1}{2} \times 0.24 \text{ m radians}$$

$$= 0.008 \text{ m radians}$$

$$\text{Total area} = 0.524 \text{ m radians}$$

$$\text{Min. area} = 0.075 \text{ m radians}$$

HEEL DUE TO WIND LOADING

A vessel heeled due to wind (*Figure 6.31*) will be in equilibrium when the righting moment $W \times GZ$ is equal and opposite to the wind heeling moment $F \times l$, when l is the vertical distance between the mid draft and the centroid of the projected side area of the vessel.

There are no statutory regulations governing the stability of vessels subject to wind heeling. However, Reference 3 makes recommendations for container vessels. These recommendations are (see *Figures 6.32* and *6.33*):

(a) The vessel is assumed to be in her worst condition.

(b) The lateral windage area is assumed to be subjected to a steady wind loading of 48.5 kg/m^2 giving the Force F.

(c) On a curve of righting moment the angle of heel due to steady wind loading is located at θ_1, where

$$\text{righting moment} = F \times l = \lambda_0$$

(d) It is then assumed that the vessel rolls 15° into the wind.

(e) Taking λ_0 as being the righting moment at θ_1, a line is drawn at $1.5\lambda_0$ parallel to the heel axis.

(f) The area S_1 between the righting moment curve and $1.5\lambda_0$ is measured.

(e) The angle θ_{dy} is found so that

$$\text{area } S_1 = \text{area } S_2$$

i.e. θ_{dy} represents the angle of heel to which the vessel would roll if the work represented by area S_1 were to be absorbed by area S_2.

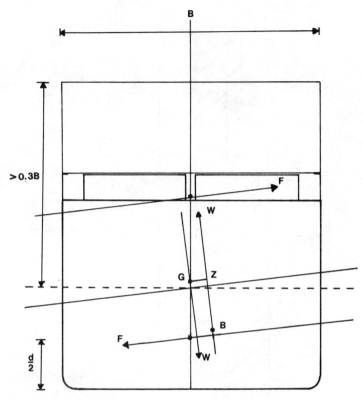

Figure 6.31 Heel due to wind loading

The instructions recommend that θ_{dy} should not exceed the angle of flooding and that

$$\theta_1 < 0.65 \text{ (angle of deck edge immersion)}$$

Example 6.5 (*Figure 6.34*)

A vessel has the following values of GZ at the angles of heel indicated

0	5	10	25	30	40
0	0.04	0.09	0.09	0.29	0.32

The vessel is displacing 32 000 tonnes, has KG, 10.3 m and KM, 10.8 m.

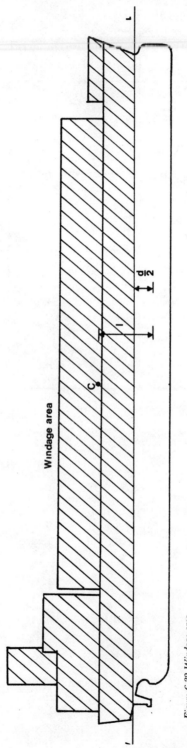

Figure 6.32 Windage area

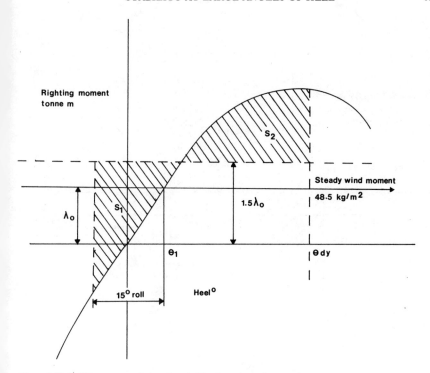

Figure 6.33 Stability recommendations for wind loading of container vessels

She is floating at draft 11 m.
The windage area is 3800 m^2 and the centroid of the area is 6 m above the waterline.
The angle of deck edge immersion is 23°.
The angle of flooding is 34°.

Assess the ability of the vessel to withstand heeling due to wind (*Figure 6.34*).

Station	GZ	Displacement	Righting moment
0	0	32 000	0
5	0.04	32 000	1280
10	0.09	32 000	2880
20	0.19	32 000	6080
30	0.29	32 000	9280
40	0.32	32 000	10 240

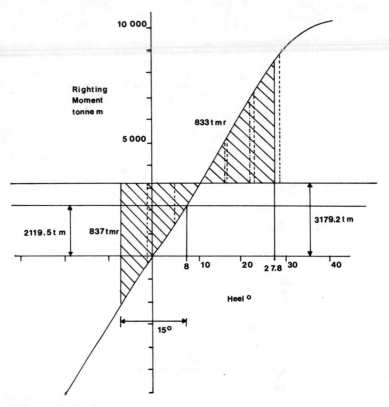

Figure 6.34 Example 6.5

$$\text{Wind lever} = \frac{d}{2} + \text{centroid windage area above } WL$$

$$= 5.5 + 6$$

$$11.5 \text{ m}$$

$$\lambda_0 = \frac{\text{windage area} \times \text{wind force} \times \text{wind level terms}}{1000}$$

$$= \frac{3800 \times 48.5 \times 11.5}{1000}$$

$$= 2119.45$$

$$1.5\lambda_0 = 3179.18$$

From curve area S_1 extend of 18°. Interval 6°

Station	Ord.	SM	F (Area)
0	5300	1	5300
1	3450	3	10 350
2	1890	3	5670
3	00	4	0
			21 320

$$S_1 = \frac{3}{8} \times h \times \sum F \text{ (Area)}$$

$$= \frac{3}{8} \times \frac{6}{57.3} \times 21\,320 \text{ tonne m radians}$$

$$S_2 = 837.1 \text{ tonne m radians}$$

Taking 18° for S_2 as first approximation

Station	Ord.	SM	F (Area)
0	0	1	0
1	2100	3	6300
2	4050	3	12 150
3	5690	1	5690
			24 140

$$S_2 = \frac{3}{8} \times h \times \sum F \text{ (Area)}$$

$$= \frac{3}{8} \times \frac{6}{57.3} \times 24\,140 \text{ tonne m radians}$$

$$= 947.9 \text{ tonne m radians}$$

$$\frac{S_1}{\text{Difference}} = \frac{837.1}{110.8}$$

Reduction in θ assuming area approximately rectangular

$$5690 \times \theta_c = 110.8$$

$$\theta_c = \frac{110.8}{5690}$$

$$= \frac{110.8 \times 57.3}{5690}$$

$$= 1.12°$$

$$\text{Range} = 18° - 1.12°$$
$$= 16.88°$$
$$\text{Interval} = 5.66°$$

Stations	Ord.	SM	F (Area)
0	0	1	0
1	1900	3	5700
2	3800	3	11 400
3	5400	1	5400
			22 500

$$S_1 = \frac{3}{8} \times \frac{5.66}{57.3} \times 22\,500 \text{ tonne m radians}$$

$$= 833.4 \text{ tonne m radians}$$

Which is sufficiently close

$$\therefore \quad \theta_{dy} = 10.8° + 16.88°$$

$$= 27.68°$$

$$\theta_f = 34° \qquad \theta_{dy} \text{ satisfactory}$$

$$\theta_1 = 8°$$

$$0.65\theta_{dc} = 0.65 \times 23°$$

$$= 14.95°$$

$$\therefore \quad \theta_1 \text{ satisfactory}$$

References

1. *Merchant Shipping Grain Rates*, (1981)
2. *IMO Safety of Life at Sea*, Chapter V, (1974)
3. *Load Line Rules Instructions to Surveyors: Appendix*

7 Calculation of righting level and assessment of stability

The methods of calculating GZ are outlined for moderate and large angles of heel. Methods of presenting stability data by reference to a single parameter are discussed.

OBJECTIVES

1. To describe the calculation of KN curves.
2. To use KN curves to find GZ for a vessel.
3. To use the wallsided formula to find GZ curves values up to deck edge immersion.
4. To use the wallsided formula to find list and loll under applicable conditions.
5. To appreciate methods of presenting simplified stability data.
6. To assess the stability of a vessel using the dead weight moment approach to simplified stability data.

KN CURVES

In *Figure 7.1* bb is a vertical line drawn through B at some angle of heel θ when the vessel is at displacement W. Then if the horizontal distance KN can be determined it can be seen that:

$$GZ = KN - KX$$
$$KX = KG \sin \theta$$
$$GZ = KN - KG \sin \theta$$

Thus if KN can be determined for a series of angles of heel say, 15°, 30°, 45°, 60°, 75° and 90°, over a range of displacement up to load

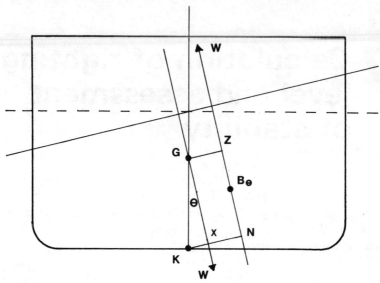

Figure 7.1 Calculation of GZ gives KN

displacement, then if KG is known it is possible to determine values of GZ and plot a GZ curve.

In principle KN can be found by taking moments of volume of the inclined vessel about a vertical axis CC (*Figure 7.2*) through K at a series of waterplanes for the angles of heel required. Then,

$$KN = \frac{\text{moment of volume about CC}}{\text{volume}}$$

The calculations are very tedious and in practice the values of KN are determined using programmes which find KN from the offsets which define the form of the vessel. The most advanced programmes take free trim into account (Chapter 6).

A set of KN curves for *MV Nonesuch* is given in *Figure 7.3*.

Example 7.1 (*Figure 7.4*)

MV Nonesuch is displacing 26 000 tonnes and has KG, 9.6 m. Does she comply with the Load Line Regulations while in this condition?

Heel	0	5	12	15	30	45	60	75	90
KN	0	1.05	2.67	3.19	6.42	8.45	8.88	8.69	7.63
$KZ \sin \theta$	0	0.84	2.00	2.48	4.80	6.79	8.31	9.27	9.6
GZ	0	0.21	0.67	0.71	1.62	1.66	0.57	−0.58	−1.97

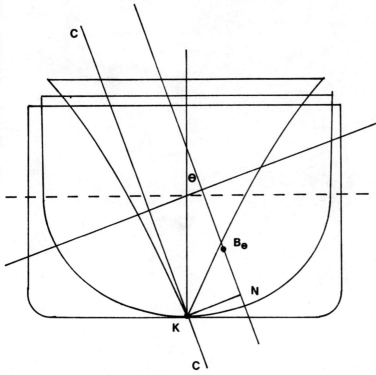

Figure 7.2 Determination of KN

KM	11.9 m
KG	9.6 m
GM	2.3 m

Max.	GZ	1.81 m	at	38°	Complies
GM		2.30 m			Complies

Heel	GZ	SM_{0-30}	$F\,(Area)_{0-30}$	SM_{0-40}	$F\,(Area)_{0-40}$
0	0	1	0.00	1	0.00
10	0.48	3	1.44	4	1.92
20	1.04	3	3.12	2	2.08
30	1.62	1	1.62	4	6.48
40	1.80			1	1.80
			6.18		12.28

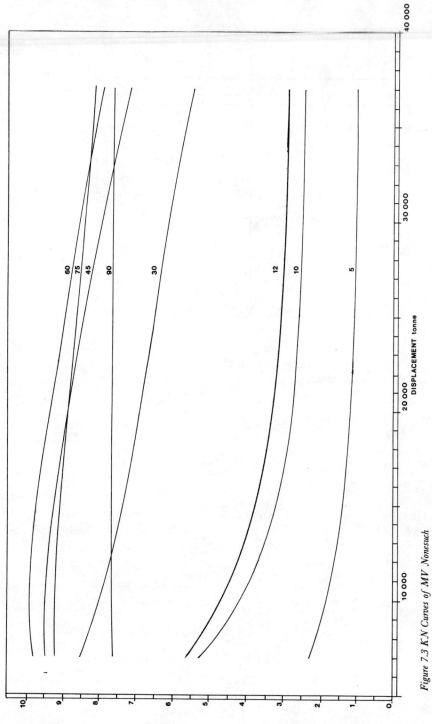

Figure 7.3 KN Curves of MV Nonesuch

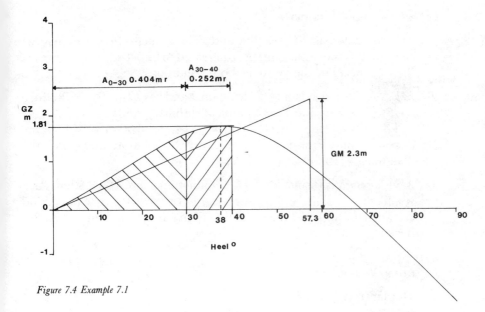

Figure 7.4 Example 7.1

$$\text{Area}_{0-30} = \frac{3}{8} \times h \times \sum F \text{ (Area)}$$

$$= \frac{3}{8} \times \frac{10}{57.3} \times 6.18 \text{ m radians}$$

$$= 0.404 \text{ m radians}$$

$$\text{Area}_{0-40} = \frac{1}{3} \times h \times \sum F \text{ (Area)}$$

$$= \frac{1}{3} \times \frac{10}{57.3} \times 12.28 \text{ m radians}$$

$$= 0.714 \text{ m radians}$$

$$\text{Area}_{30-40} = 0.252 \text{ m radians}$$

$$\text{Area}_{0-30} = 0.404 \text{ m radians} \qquad \text{Complies}$$

$$0-40 = 0.714 \text{ m radians} \qquad \text{Complies}$$

$$30-40 = 0.252 \text{ m radians} \qquad \text{Complies}$$

Questions on GZ curves

1   *MV Nonesuch* has KG, 12.0 m and displacement 14 000 tonnes, at constant KG the displacement is increased to 24 000 tonnes. Compare the GZ curves of the vessel in both conditions.

2   *MV Nonesuch* is displacing 30 000 tonnes and has KG, 8.00 m. Produce a GZ curve and find the dynamical stability at 40° heel.

   The KG of the vessel is then raised to 8.50 m at the same displacement. Produce the amended GZ curve and find the dynamical stability of the vessel at 40° heel.

3   *MV Nonesuch* is displacing 24 000 tonnes and has KG, 8.50 m. As a result of cargo shifting the centre of gravity moves 0.5 m to port. Find the angle of heel and the reduction of dynamical stability up to 30° heel.

GZ curves (answers)

1   Displacement 14 000
   GM 4.10 m

Heel	0	5	15	30	45	60	75	90
GZ m	0	0.38	1.12	1.42	10.93	−0.65		

   Displacement 24 000
   GM 0.000 m

Heel	0	5	15	30	45	60
GZ m	0.0	0.0	0.12	0.44	0.05	−1.50

2   (i) GM 3.560 m. Dynamical stability 0.93 m radians

Heel	0	5	15	30	45	60	75	90
GZ	0	0.35	0.98	2.10	2.07	1.60	0.75	−0.48

   (ii) GM 3.060. Dynamical stability 0.12 m radians

Heel	0	5	15	30	45	60	75	90
GZ	0	0.31	0.85	1.85	1.72	1.17	0.27	−0.98

3   Initial Condition GM 3.505 m. Dynamical Stability 30° 0.539 m radians

Heel	0	5	15	30	45	60	75	90
GZ	0	0.29	0.95	2.38	2.55	1.65	0.46	−0.92

   Find Condition GM 3.505. Dynamical Stability 30° 0.357 m radians

Heel	0	5	15	30	45	60	75	90
GZ	−0.5	−0.21	0.47	1.95	2.20	1.40	0.33	−0.92

WALL-SIDED VESSELS

If it is assumed that the sides of the vessel are parallel, then it is possible to calculate values of GZ for the early part of the GZ curve. Under these circumstances, the 'Wall Sided Formula' will give accurate results so long as successive waterplanes intersect at the centreline.

In *Figure 7.5* a vessel breadth B, length L, is heeled to an angle of heel θ by an external force. The centre of buoyancy will move from B_0 to B_1 and vertically from B_1 to B_2.

$$GZ = GX + XZ$$

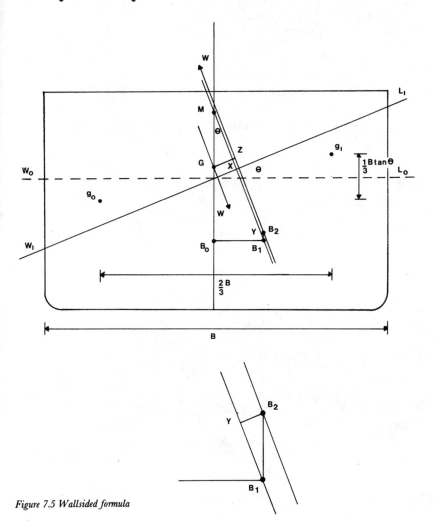

Figure 7.5 Wallsided formula

The component GX is due to the horizontal shift of the centre of buoyancy from B_0 to B_1. The component XZ is due to the vertical shift of the centre buoyancy from B_1 to B_2.

Now if a wedge of buoyancy volume V has moved across the vessel when she is heeled to some angle and the centre of buoyancy of the wedge has moved from g_0 to g_1 then considering the horizontal shift

$$B_0 B_1 = \frac{V \times (g_0 g_1)_\text{H}}{\nabla}$$

$$V \times (g_0 g_1)_\text{H} = \int_0^L \frac{1}{2} \times \frac{B}{2} \times \tan\theta \times dl \times \frac{2}{3} \times \frac{B}{2} \times 2$$

$$= \tan\theta \int_0^L \frac{B^3 \times dl}{12}$$

$$= I \times \tan\theta$$

(See derivation of BM, Chapter 4)

$$B_0 B_1 = \frac{I}{\nabla} \tan\theta$$

$$B_0 B_1 = B_0 M \tan\theta$$

$\therefore$ the vertical line through B_1 intersects the centre line at M, the upright metacentre.

$$\therefore \quad GX = GM_0 \sin\theta$$

Considering the vertical shift $B_1 B_2$

$$B_1 B_2 = \frac{V \times (g_0 g_1)_\text{v}}{\nabla}$$

$$V \times (g_0 g_1)_\text{v} = \int_0^L \frac{B^2}{8} \times \tan\theta \times dl \times \frac{1}{3} \times \frac{B}{2} \times \tan\theta \times 2$$

$$= \frac{\tan^2\theta}{2} \int_0^L \frac{B^3 \times dl}{12}$$

$$= \frac{\tan^2\theta}{2} I$$

$$B_1B_2 = \frac{1}{2} \times \frac{I}{V} \times \tan^2 \theta$$

$$B_1B_2 = \frac{1}{2} \times B_0M \times \tan^2 \theta$$

$$B_2Y = \frac{1}{2} \times B_0M \times \tan^2 \theta \times \sin \theta$$

$$= XZ$$

$$\therefore \quad GZ = GM \sin \theta + \frac{1}{2} B_0M \tan^2 \theta \sin \theta$$

$$GZ = \sin \theta \left(GM + \frac{1}{2} B_0M \tan^2 \theta \right)$$

This formula gives an analytical base to the description of the early part of the GZ curve given in Chapter 6. $GM \sin \theta$ represents the small angle formula for GZ.

$\frac{1}{2}B_0M \tan^2 \theta \sin \theta$ represents the additional shift of B due to larger wedges of buoyancy being moved across the vessel. It should also be noted that when θ is small:

$$\tfrac{1}{2}B_0M \tan^2 \theta \sin \theta = \frac{1}{2} B_0M\theta_c{}^3$$

and therefore a small quantity. This sustains the small angle assumption

$$GZ = G_0M \sin \theta$$

Within the limits of the assumptions, used to formulate the wall sided formula, the formula can be employed to find the angle of loll and the list which results when weights are shifted when GM is zero.

ANGLE OF LOLL (*Figure 7.6*)

If the vessel initially has a negative metacentric height GM_0.

As the vessel heels she may reach some angle of heel θ when B_L and G are once more in the same vertical line and the metacentre is at M_L above G.

At this angle of heel

$$GZ = 0$$

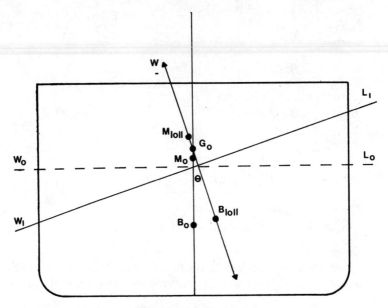

Figure 7.6 Determination of angle of loll using wall-sided formula

Using the wall sided formula

$$GZ = \sin \theta \left(GM_0 + \frac{1}{2} BM \tan^2 \theta \right)$$

$$0 = \sin \theta \left(GM_0 + \frac{1}{2} BM \tan^2 \theta \right)$$

$\therefore$ either $\sin \theta = 0$

which can only occur if the vessel is upright or

$$GM_0 + \frac{1}{2} \times B_0 M \times \tan^2 \theta = 0$$

$\therefore$ $\quad \dfrac{1}{2} \times B_0 M \times \tan^2 \theta = - GM_0$

$$\tan^2 = \frac{-2 \times GM_0}{B_0 M}$$

$$\tan \theta = \sqrt{\frac{-2 \times GM_0}{B_0 M}}$$

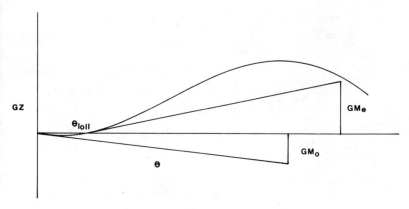

Figure 7.7 Determination of GM lolled

Note that the GM is the initial GM and since the vessel is initially unstable this GM is negative. Thus the numerator in the above expression will be positive. Also this expression will always give a value for the angle of loll. However, the formula is limited by the same factors as those for the wall sided formula.

Thus, if the angle of loll is greater than the angle of deck edge immersion, the formula is not valid and whether or not the vessel reaches an angle of loll can only be found by constructing a GZ curve.

In *Figure 7.7* it can be seen that at the angle of loll, the vessel has a positive GM_L. If it is recalled that GM can be found by measuring the slope of the GZ curve where it crosses the heel axis (Chapter 6), then the GM at the angle of loll can be found (*Figure 7.7*).

$$GZ = \sin \theta \left(GZ + \frac{1}{2} BM_0 \tan^2 \theta \right)$$

Differentiating using the product rule

$$\frac{d(GZ)}{d\theta} = \cos \theta \left(GM_0 + \frac{1}{2} BM_0 \tan^2 \theta \right)$$

$$+ \sin \theta BM_0 \tan \theta \sec \theta$$

Putting $\theta = \theta_1$

$$\tan^2 \theta_1 = \frac{-2GM_0}{BM_0}$$

since θ_1 is the angle of loll

$$\frac{d(GZ)}{d\theta} = \cos\theta\left(GM_0 + \frac{1}{2}BM_0\left(\frac{-2GM_0}{BM_0}\right)\right)$$
$$+ BM_0\left(\frac{-2GM_0}{BM_0}\right)\sec\theta_1$$

$$GM_{\text{loll}} = \frac{d(GZ)}{d\theta_1}$$

$$GM_{\text{loll}} = 0 - 2GM_0\sec\theta_1$$

$$GM_{\text{loll}} = -2GM_0\sec\theta_1$$

LIST WHEN $GM_0 = 0$

When $GM_0 = 0$ the relationship $\tan\theta = G_0G_1/G_1M_0$ would give the result that $\theta = 90°$.

However, in *Figure 7.8* using the wall sided formula

$$GZ = \sin\theta\left(GM + \frac{1}{2}BM\tan^2\theta\right)$$

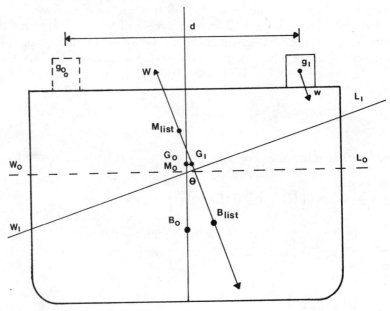

Figure 7.8 List when $G_0M_0 = 0$

and putting

$$GZ = G_0G_1 \cos \theta$$

$$= \frac{W \times d}{W} \cos \theta$$

$$GM = 0$$

$$\frac{W \times d}{W} \times \cos = \frac{1}{2} \times BM_0 \times \tan^2 \theta \times \sin \theta$$

$$\frac{w \times d}{W} = \frac{1}{2} \times BM_0 \times \tan^3 \theta$$

$$\tan^3 \theta = \frac{2 \times w \times d}{BM_0 W}$$

$$\tan \theta = 3 \sqrt{\frac{2 \times w \times d}{BM_0 W}}$$

Again it should be noted that this relationship only holds good within the wall sided limits. Any case which falls outside those limits must be examined by constructing a GZ curve and plotting the $G_0G_1 \cos \theta$ curve on it.

Example 7.2

MV Nonesuch has draft 8 m and *KG*, 10 m. Compare values from *KN* curves with values from the wall sided formula for 5°, 12° and 15° heel.

From Hydrostatic data

KM	11.6 m	KM	11.6 m	Displacement 28 200 tonnes	
KG	10.0 m	KB	4.2 m		
GM	1.6 m	BM	7.4 m		

$$GZ = \sin \theta \left(GM + \frac{1}{2} BM \tan^2 \theta \right)$$

$$= \sin \theta (1.6 + 3.7 \tan^2 \theta)$$

Heel	GZ
5°	0.140 m
12°	0.367 m
15°	0.483 m

From KN curves

	KN	KG sin θ	GZ m
5°	1.02	0.87	0.15
12°	2.49	2.08	0.41
15°	3.08 m	2.59	0.49

Example 7.7

A box-shaped vessel has length 8.0 m, breadth 9 m, depth 8.5 m and is floating at draft 5 m with KG, 3.7 m in salt water. Find the angle of loll and the GM_{loll} if a weight of 500 tonnes is loaded at KG, 8 m.

$$\text{Initial displacement} = L \times B \times d \times \rho$$
$$= 80 \times 90 \times 5 \times 1.025 \text{ tonnes}$$
$$= 3690 \text{ tonnes}$$

Load KG

Weight	KG	Moment
3690	3.7	13 653
500	8.0	4 000
4190		17 653

$$KG = \frac{\text{moment}}{\text{weight}}$$

$$= \frac{17\,653}{4190} \text{ m}$$

$$= 4.213 \text{ m}$$

Load draft

$$d = \frac{\text{displacement}}{L \times B \times \rho}$$

$$= \frac{4190}{80 \times 9 \times 1.025} \text{ m}$$

$$= 5.678 \text{ m}$$

Final KM

$$= \frac{d}{2} + \frac{B^2}{12d}$$

$$= \frac{5.678}{2} + \frac{9^2}{12 \times 5.678}$$

$KM = 4.028$ m

$KG = 4.213$ m

$GM = -0.185$ m

$$\tan \text{loll} = \sqrt{\frac{-2GM}{BM}}$$

$$= \sqrt{\frac{-2 \times (-0.185)}{1.189}}$$

$\text{loll} = 29.15°$

$GM_{\text{loll}} = -2 \times GM_0 \sec \text{loll}$

$\qquad = -2 \times 0.185 \times \sec 29.15°$

$\qquad = 0.424$ m

Example 7.8

A vessel has displacement 25 000 tonnes KG 10.6 m, KM 12.0 m, KB 6.1 m. Angle of deck edge immersion 27°. Estimate the dynamical stability at 20° heel

$GM = KM - KG$

$\qquad = 12.0 - 10.6 = 1.4$ m

$BM = KM - KB$

$\qquad = 12.0 - 6.1 = 5.9$ m

$$\frac{BM}{2} = 2.95 \text{ m}$$

Calculating GZ at 5° intervals is

$$GZ = \sin \theta \left(GM + \frac{1}{2} BM \tan^2 \theta \right)$$

Heel	0	5	10	15	20	
$2.95 \tan^2 \theta$	0.000	0.023	0.092	0.212	0.391	
$+1.4$	1.400	1.400	1.400	1.400	1.400	
	1.400	1.423	1.492	1.612	1.791	
$\times \sin \theta$	0.000	0.098	0.174	0.259	0.342	
GZ	0.000	0.124	0.260	0.418	0.613	
SM	1	4	2	4	1	
F (Area)	0.000	0.496	0.520	1.672	0.613	$= 3.301$

$$\text{Area} = \frac{h}{3} \Sigma F \text{ (Area)}$$

$$= \frac{1}{3} \times \frac{5}{57.3} \times 3.301 \text{ m radians}$$

$$= 0.096 \text{ m radians}$$

$$\text{Dynamical Stability} = \text{Area} \times W$$

$$= 0.096 \times 25\,000 \text{ tonnes m}$$

$$= 2400.0 \text{ tonne m}$$

Example 7.9

A box shaped vessel has length 120 m, breadth 18 m, depth 12 m is floating at draft 8 m in fresh water with KG 7.278 m. A weight of 432 tonnes is loaded on deck at Kg 12 m. Find the dynamical stability of the vessel up to the angle of deck edge immersion.

$$\text{Displacement} = L \times B \times d \times \rho$$

$$= 120 \times 18 \times 8 \times 1 \text{ tonne}$$

$$= 17\,280 \text{ tonne}$$

Weight	KG	Moment
17 280	7 278	125 755.2
432	12 000	5 184.0
17 712		130 939.2

$$KG = \frac{\text{moment}}{\text{weight}}$$

$$= \frac{130\,939.2}{17\,712} = 7.393 \text{ m}$$

$$\text{Draft} = \frac{\text{displacement}}{L \times B \times \rho}$$

$$= \frac{17\,712}{120 \times 18 \times 1} = 8.2 \text{ m}$$

$$KM = \frac{d}{2} + \frac{B^2}{12d}$$

$$= \frac{8.2}{2} \text{ m} + \frac{18^2}{12 \times 8.2} \text{ m}$$

$$KM = 7.393 \text{ m}$$
$$KG = 7.393 \text{ m}$$
$$\overline{}$$
$$GM = 0.000 \text{ m}$$
$$\overline{}$$

Angle of deck edge immersion

$$\tan \theta = \frac{\text{freeboard}}{\text{half breadth}}$$

$$= \frac{12 - 8.2}{9}$$

$$\theta = 22.89°$$

$$GZ = \sin \theta \left(GM + \frac{1}{2} BM \tan^2 \theta \right)$$

$$GZ = \frac{1}{2} BM \tan^2 \theta \sin \theta$$

$$BM = 3.293 \text{ m}$$

$$\frac{BM}{2} = 1.646 \text{ m}$$

$$GZ = 1.646 \tan^2 \theta \sin \theta$$

Area 0–20

Heel	GZ	SM	F (Area)
0	0.000	1	0.000
5	0.001	4	0.004
10	0.009	2	0.018
15	0.031	4	0.124
20	0.075	1	0.075
			0.221

$$\text{Area}_{0-20} = \frac{1}{3} \times \frac{5}{57.3} \times 0.221$$

$$= 0.0064 \text{ m radians}$$

Area$_{20-22.89}$

Heel	GZ
20	0.075
22.89	0.114
	0.189

Using trapezoid rule

$$\text{Area} = \frac{1}{2} \times \frac{2.89}{57.3} \times 0.189$$

$$= 0.0048 \text{ m radians}$$

$$\text{Area}_{0-20} = 0.0064 \text{ m radians}$$
$$\text{Area}_{0-22.9} = 0.0112 \text{ m radians}$$

Dynamical stability = area × displacement

$$= 0.0112 \times 17\,712 \text{ tonne m}$$

$$= 198.4 \text{ tonne m}$$

Example 7.10

A vessel has displacement 22 500 tonnes, *KG* 7.3 m, *KM* 7.4 m, *BM* 3.6 m. A heavy lift of 150 tonnes at *KG* 2.5 m on the centreline is lifted using a derrick with head 17.5 m above the keel and swung 3 m to port.

Determine list

Weight	KG	Moment	G from CL centreline	Moment P
22 500	7.3	164 250		
− 150	2.5	− 375		
+ 150	17.5	2 625	3	450
22 500		166 500		450

$$KG = \frac{\text{moment}}{\text{weight}} \qquad\qquad G_0G_1 = \frac{\text{moment}}{\text{weight}}$$

$$= \frac{166\,500}{22\,500} \qquad\qquad\qquad = \frac{450}{22\,500}$$

$$= 7.400 \text{ m} \qquad\qquad\qquad = 0.02 \text{ m}$$

$$KM = 7.400 \text{ m}$$

$$GM = 0.000 \text{ m}$$

$$\tan \text{list} = 3\sqrt{\frac{2G_0G_1}{BM}} = 3\sqrt{\frac{2 \times 0.02}{3.6}}$$

$$\tan \text{list} = 0.223$$

$$\text{list} = 12.57° \text{ port}$$

QUESTIONS ON WALL SIDED FORMULA

1   A box shaped vessel has length 100 m, breadth 10 m, depth 6 m and is floating at a draft of 4 m, KG 3.5 m. Find the dynamical stability of the vessel at 20° heel.

2   A vessel has KG 7.68 m and KM 7.60 m. She is at an angle of inclination of 11° to port and is displacing 7500 tonnes. What will be the list if a port side double bottom tank is filled with 200 tonnes of water with a Kg of 0.5 m, 4 m to port of the centreline? Find the GZ of the vessel at 20° heel after filling the double bottom tank.

3   A vessel displacement 22 500 tonnes, KG 7.3 m, KM 7.4 m, BM 3.6 m. A lift of 150 tonnes at Kg 2.5 m on the centreline is lifted using a derrick with head 17.5 m above the keel and swung 3 m to port. Find the list.

4   A box shaped vessel has length 80 m, breadth 9 m, depth 8.5 m and is floating at draft 5 m with KG 3.7 m in salt water. Find the loll and the GM lolled if a weight of 500 tonnes is loaded at Kg 8 m.

WALL SIDED FORMULA (ANSWERS)

1   $GM = 0.583$ m
    Area 0–$20 = 0.0392$ m radians
    Dynamical Stability 160.72 tonne m.

2   $BM = 4.235$ m
    Using WSF and $G_0G_1 \cos \theta$ curve to find list
    list $= 17°$ to port
    GZ at $20° = 0.0342$ m

3   List 12.57° p.

4   Loll 29.15°
    GM lolled 0.424 m.

SIMPLIFIED STABILITY INFORMATION

The object of this type of information is to enable the stability of the vessel to be assessed with respect to a single parameter such as KG, GM or more commonly, deadweight moment.

If a boundary line between deficient and adequate stability can be established such that on the adequate side of the line the vessel complies with, say, the Load Line Rules, then the calculation of all aspects of stability can be avoided. *Figure 7.9* shows sketches of possible forms of this information. The method of use is similar for all types of presentations. We will concentrate on the deadweight moment method.

Deadweight moment is the moment of cargo, fresh water, fuel stores, etc. about the keel, including free surface moment, i.e.:

$$KG = \frac{\text{moment of weight about keel}}{\text{displacement}}$$

$$KG = \frac{\text{light ship moment} + \text{deadweight moment}}{\text{displacement}}$$

Thus if deadweight moment is increased for a particular displacement, KG will be increased and stability reduced.

Figure 7.10 is a possible deadweight moment curve for the vessel used for the hydrostatic data. Note that at light displacement the vessel will not have zero maximum permissable deadweight moment. This is because the vessel will always have an adequate reserve of stability in the light condition and will have a substantial permissable deadweight moment in this condition.

The use of the curve is best illustrated by an example.

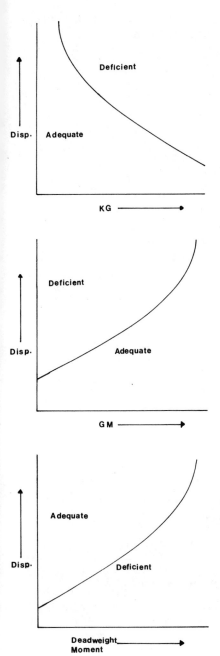

Figure 7.9 *Methods of presenting simplified stability data*

Example 7.11

MV Nonesuch (Figure 7.10) is displacing 15 000 tonnes and has deadweight moment 40 000 tonnes m. She loads cargo as follows:

    No 1   4000 tonnes Kg 7 m
    No 2   3000 tonnes Kg 6.5 m
    No 3   4500 tonnes Kg 7.2 m
    No 4   3000 tonnes Kg 7.3 m
    No 5   4000 tonnes Kg 6.9 m

She discharges

    No 1   TSWT 800 tonnes Kg 12 m
    No 5   TSWT 600 tonnes Kg 11.8 m

During the subsequent voyage, it is estimated that she will use 300 tonnes of fuel from Kg 2.0 m introducing a free surface moment of 800 tonnes m. Will she be in a satisfactory condition on arrival?

Compt	Weight	Kg	Moment	FSM
1	15 000		40 000	
1	4 000	7.0	28 000	
2	3 000	6.5	19 500	
3	4 500	7.2	32 400	
4	3 000	7.3	21 900	
5	4 000	6.9	27 600	
1 TSWT	− 800	12.0	− 9 600	
5 TSWT	− 600	11.8	− 7 080	
Fuel	− 300	2.0	− 600	+ 800
	31 800		152 120	
			800	
			152 920	

Plotting displacement   31 800 tonnes
Deadweight moment 152 920 tonnes m

Indicates that the vessel is in a satisfactory condition.

Effect loading or discharging from a particular Kg

Suppose in *Figure 7.11* a vessel is loaded in the condition indicated by C_0 then if cargo is loaded anywhere in the vessel

(a) Displacement must increase;
(b) Deadweight moment must increase.

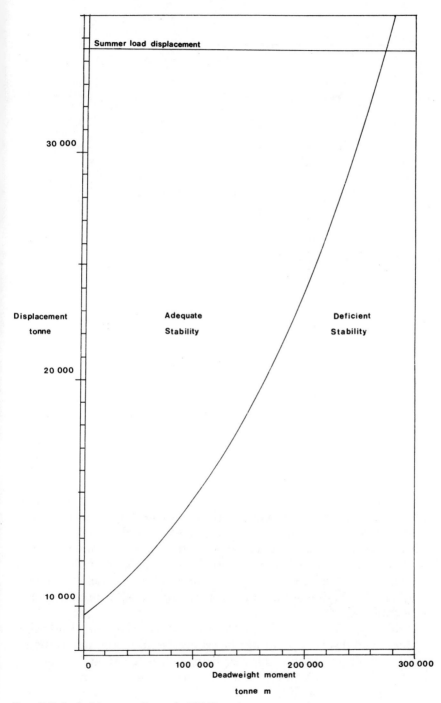

Figure 7.10 Deadweight moment diagram for MV Nonesuch

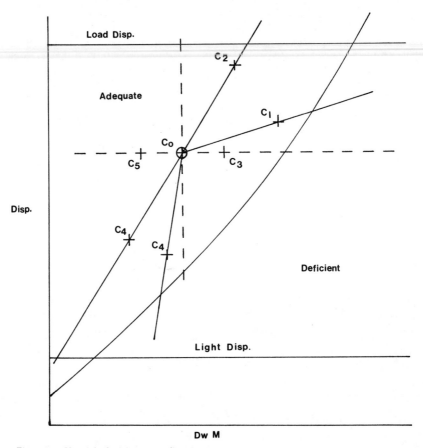

Figure 7.11 Use of deadweight moment diagram

The vessel could move to condition C_1 such that if cargo continued to be loaded in that position, the vessel's stability would become deficient. Alternatively, the vessel could move towards condition C_2 where, if cargo continued to be loaded, the amount loaded would be limited by the summer load displacement. It is possible for the vessel to move to condition C_3, if a free surface moment is increased while displacement is kept constant or weight is redistributed.

Similarly, if cargo is discharged from a particular position, deadweight moment and displacement must be reduced to some conditions such as C_4. The limiting factor in this case being deficient stability or reaching light displacement. A vessel could move to condition C_5 by reducing free surface moment or redistributing weight.

Example 7.12

MV Nonesuch has displacement 25 000 tonnes and deadweight moment 180 000 tonnes m. How much deck cargo may she load at *Kg* 14 m if the vessel is to be in a satisfactory stability condition after loading the cargo.

Try loading 2000 tonnes at *Kg* 14 m.

Weight	Kg	Moment
25 000		180 000
2 000	14	28 000
27 000		208 000

Plot points
Displacement 25 000   DWM 180 000
Displacement 27 000   DWM 208 000
on data sheet.

Join the points and produce to cut the boundary line between adequate and deficient stability.

Read displacement at crossing point	= 27 700 tonnes
Present displacement	= 25 000 tonnes
Cargo to load	2 700 tonnes

QUESTIONS ON USE OF SIMPLIFIED STABILITY DATA

1   *MV Nonesuch* has displacement 14 000 tonnes and deadweight moment 40 000 tonnes m.

She loads:

No 1		5000 tonnes	Kg 6.8 m	
No 3		7000 tonnes	Kg 7.2 m	
No 5		6000 tonnes	Kg 8.3 m	
No 2 TSWT		1000 tonnes	Kg 9.0 m	FSM 600 tonnes m
No 4 TSWT		900 tonnes	Kg 9.2 m	FSM 580 tonnes m

Does the vessel have adequate stability on completion.

2   *MV Nonesuch* has displacement 31 000 tonnes and deadweight moment 240 000 tonnes m on departure.

During the voyage it is estimated that

(i)   Deck cargo will absorb 200 tonnes of water at Kg, 12 m.
(ii) The vessel will use 400 tonnes of fuel from Kg, 0.7 m.
(iii) Using the fuel will remove a free surface moment of 800 tonnes m.

Find the deadweight moment on arrival and whether the vessel will be in a satisfactory condition or not.

3 *MV Nonesuch* has displacement 17 000 tonnes and deadweight moment 140 000 tonnes m.

She loads
5000 tonnes Kg 8.0 m
7000 tonnes Kg 7.0 m
 250 tonnes Kg 1.2 m   FSM 320 tonnes m.

How much cargo may she load at Kg, 11.0 m in order to be in a satisfactory stability condition.

4 *MV Nonesuch* has displacement 20 000 tonnes and deadweight moment 160 000 tonnes m. She is to load a total of 5000 tonnes of ballast.

(i) 1000 tonnes of ballast Kg, 1.0 m. FSM 0 on completion.
FSM on commencing ballasting this tank 400 tonnes m.
(ii) 4000 tonnes of ballast Kg, 7.0 m. FSM 300 tonnes m on completion.
While ballasting this compartment, the worst conditions will occur when there are 2800 tonnes of ballast on board with a Kg of 5.0 m and a FSM of 10 000 tonnes m.

Ballast the vessel so that she is always in a satisfactory condition, stating the displacement and deadweight moments at each stage.

SIMPLIFIED STABILITY DATA (ANSWERS)

1 Displacement 33 900   DWM 192 660
   Adequate

2 Displacement 30 800   DWM 242 320
   Adequate

3 Load 1250 tonnes more at Kg 11 m.

4 Loading 1000 tonnes

Initial	Disp. 20 000	DWM 160 400	Marginal
Final	Disp. 21 000	DWM 161 000	Marginal
Loading	4000 tonnes		
Worst	Disp. 23 800	DWM 185 000	Adequate
Final	Disp. 25 000	DWM 189 300	Adequate

8 Longitudinal stability—trim

This chapter examines the conditions which can cause a ship to change trim due to shifting, loading and discharging weights. Change of trim due to change of density is considered. The determination of displacement in a trimmed sagged, or hogged ship is introduced.

OBJECTIVES

1. Definition of longitudinal metacentre and centre of flotation.
2. Determination of moment to change trim one centimetre (MCTC) and the position of the centre of flotation.
3. Determination of drafts of the ship after handling moderate and large weights.
4. Loading and discharging to a given draft.
5. Determination of displacement (Draft Survey).

Definition of trim

Trim is the difference between the forward and after drafts of a vessel, by convention trim is measured in centimetres.

If a vessel is initially floating at waterline W_0L_0 (*Figure 8.1*) and changes trim at constant displacement to float at W_1L_1, then the difference between the trim at waterlines W_0L_0 and W_1L_1 is the change in trim. In practice the difference between trim and change of trim is obvious, however, students should take care to distinguish between these values when carrying out calculations when studying.

The two waterlines in *Figure 8.1* intersect at the centre of flotation (F). Since the displacement waterline W_0L_0 and W_1L_1 is constant the wedges of buoyance W_0, F, W_1 and L_0, F, L_1 are equal. This can only be the case if the areas of water plane forward and aft of F are equal so that the

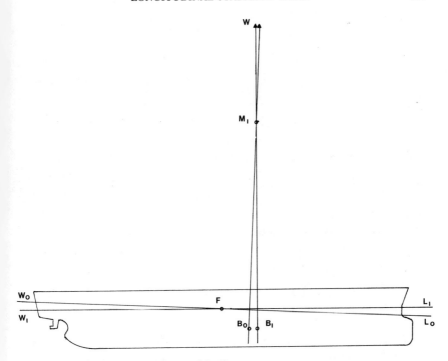

Figure 8.1 Longitudinal metacentre and centre of flotation

volumes swept out as the vessels changes trim by a moderate amount are equal. Hence, the centre of flotation is at the centroid of the water plane.

In *Figure 8.1* the centre of buoyancy has moved from B_0 to B_1, the displacement vectors through B_0 and B_1 intersect at M_L the longitudinal metacentre. Using similar methods used in Chapter 4 it can be shown that

$$B_0 M_L = \frac{I_L}{V}$$

where I_L is the second moment of area of the waterplane about a transverse axis through F.

All normal ship shapes are very stable longitudinally, so in considering longitudinal stability we are only concerned with the changes of trim which occur when the centres of buoyancy and gravity are moved out of the same vertical line. The ship will always trim until the centre of gravity and centre of buoyancy are in the same vertical line.

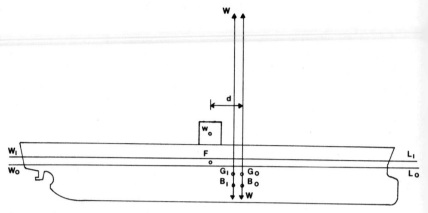

Figure 8.2 Loading a weight at centre of flotation

EFFECT OF LOADING OR DISCHARGING A WEIGHT AT THE CENTRE OF FLOTATION

In *Figure 8.2* on a vessel displacement W, a weight w is loaded directly above F the centre of flotation a horizontal distance dm from the centre of gravity G_0 and the centre of buoyancy B_0. The vessel will sink from waterline W_0L_0 to W_1L_1, a layer of buoyancy volume v is added. Provided the sinkage is small the centroid of this added volume will be at F.

Then the shift of Centre of Gravity is G_0G_1

$$G_0G_1 = \frac{w \times d}{W + w}$$

and the shift of the centre of buoyancy B_0B_1 is

$$B_0B_1 = \frac{v \times d}{V + v}\frac{\rho}{\rho} = \frac{w \times d}{W + w}$$

Hence

$$B_0B_1 = G_0G_1$$

The vessel does not change trim if a moderate weight is loaded in line with F.

CHANGE OF TRIM DUE TO SHIFTING A MODERATE WEIGHT ALREADY ON BOARD

In *Figure 8.3* the vessel displacement W, length L is floating at waterline W_0L_0 with centre of gravity at B_0 centre of gravity at G_0 and centred flotation at F.

A moderate weight w already on board is moved a distance dm from forward to aft.

The Centre of Gravity of the vessel will move from G_0 to G_1 and the vessel will trim about F until the centre of buoyancy has moved to B_1 vertically beneath G_1. The vessel will now be at waterline W_1L_1.

W_0C is a construction line drawn parallel to W_1L_1, M_L in the longitudinal metacentre.

$$\text{Change of trim} = W_0W_1 + L_0L_1$$

$$W_0W_1 = L_1C$$

$$\text{Change of trim} = L_0L_1 + L_1C$$

$$= L_0C$$

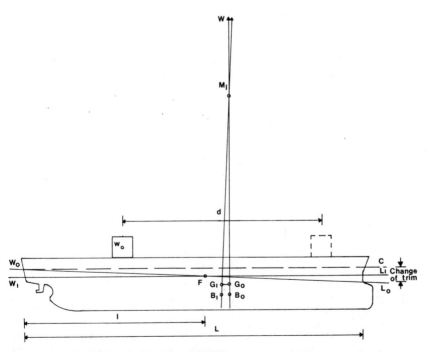

Figure 8.3 Change of trim and determination of MCTC

In triangles

$$W_0 L_0 C, \qquad\qquad M_L G_0 G_1$$

$$\angle C = \angle G_0 \simeq 90°$$

$$\angle W_0 = \angle M_L = \text{trim angle}$$

$\therefore \quad \triangle W_0 L_0 C$ is similar to $\triangle M_L G_0 G_1$

$$\therefore \quad \frac{L_0 C}{W_0 L_0} = \frac{G_0 G_1}{G_0 M_L}$$

$$L_0 C = W_0 L_0 \times \frac{G_0 G_1}{G_0 M_L}$$

$$G_0 G_1 = \frac{w \times d}{W}$$

$$L_0 C = \text{Change of trim}$$

$$W_0 L_0 = \frac{L \times w \times d}{W \times G_0 M_L}$$

$$\text{Change of trim} = \frac{L \times w \times d}{W \times G_0 M_L}$$

This formula is not very convenient to use and in practice, for moderate weights, it is better to use the Moment to Change Trim One Centimetre ($MCTC$).

The $MCTC$ is the value of $w \times d$ which will change trim one centimetre. Noting that the units of the formula above and

$$\frac{m \times \text{tonne} \times m}{\text{tonne} \times m} = m$$

Then $w \times d$ is the $MCTC$

$$\frac{1}{100} = \frac{L \times MCTC}{W \times G_0 M_L}$$

$$MCTC = \frac{W \times G_0 M_L}{100 \times L}$$

This value of $MCTC$ is given as part of the hydrostatic data of the vessel. The value of $G_0 M_L$ will vary with the loading condition of the ship, in an unpredictable manner and is thus not very useful for calculating hydrostatic data. However, for ship shapes, the difference between $B_0 M_L$

and G_0M_L will be small when compared to the value of B_0M_L. B_0M_L can be calculated for each displacement. Therefore to a good approximation

$$MCTC = \frac{W \times B_0M_L}{100 \times L}$$

For moderate weights

$$\text{Change of trim} = \frac{\text{moment changing trim}}{MCTC}$$

Distribution of trim

If, in *Figure 8.4*, the centre of flotation is a distance 1 m from the after perpendicular in a vessel length L, and the vessel changes trim through t cm then

$$\text{Change in trim after} = \frac{l}{L} \times t$$

$$\text{Change in trim forward} = \frac{(L-l)}{L} \times t$$

If weights are loaded or discharged then the change in draft has to be taken into account from

$$\text{Sinkage} = \frac{w}{TPC}$$

Example 8.1 (*Figure 8.5*)

A vessel displacing 30 000 tonnes is floating at drafts F 8.3 m, A 9.6 m. *MCTC*, 300 tonne m/cm. Centre of Flotation, 109 m forward of after perpendicular (AP), length, 210 m.

Find the drafts fore and aft if 1000 tonnes of ballast are moved from a

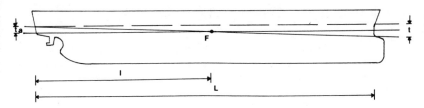

Figure 8.4 Distribution of trim

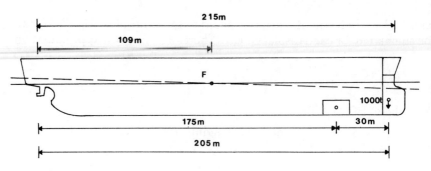

Figure 8.5 Example 8.1

tank centre of gravity 175 m forward of AP to a tank 205 m forward of AP.

$$\text{Change of trim} = \frac{\text{moment changing trim}}{MCTC}$$

$$= \frac{1000 \times 30}{300} = 100 \text{ cm}$$

$$\text{Change of aft} = \frac{l}{L} \times \text{Change of trim}$$

$$= \frac{109}{210} \times 100 \text{ cm} = -51.9$$

Change in trim forward = + 48.1 cm

F	A
8.300 m	9.600 m
+ 0.481 m	− 0.519 m
8.781 m	9.081 m

Draft Forward 8.78 m   Aft 9.08 m

Example 8.2 *(Figure 8.6)*

A vessel floating at draft Forward 9.84 m; Aft 10.62 m.

She loads

	Weight (tonne)	LCG from AP (m)
	450	25
	320	100
Discharges	140	110

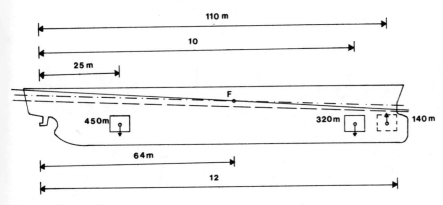

Figure 8.6 Example 8.2

TPC, 26 tonne/cm; MCTC, 148 tonne m/cm; LCF, 64 m forward of AP; length, 120 m.

Weight tonne	LCG from F	Moments forward tonne m	Moments aft tonne m
450	39		17 550
320	36	11 520	
−140	46	−6 440	
630		5 080	17 550
			5 080
			12 470 by stern

$$\text{Change of trim} = \frac{\text{moment changing trim}}{MCTC}$$

$$= \frac{12\,470}{148} = 84.3 \text{ cm}$$

$$\text{Change of trim after} = \frac{l}{L} \times \text{change of trim}$$

$$= \frac{64}{120} \times 84.3 = 45 \text{ cm}$$

Change of trim forward $= 39.3$ cm

$$\text{Sinkage} - \frac{w}{TPC}$$

$$= \frac{639}{26} = 24.2$$

	F	A
Initial draft	9.840	10.620
Sinkage	0.242	0.242
	10.082	10.862
Trim	− 0.393	+ 0.450
Final draft	9.689 m	11.312 m

Example 8.3

A vessel about to enter port has drafts: forward, 11.20 m; aft, 12.00 m. If the vessel is to enter port on an even keel find the amount of water ballast to transfer from the double bottom LCG 80 m forward to AP to the fore peak tank 195 m forward of AP.

MCTC, 210 tonne m/cm; LCF, 95 m forward of AP; length, 200 m.

Let w be weight of ballast to shifted a distance of m

$$d = 195 \text{ m} - 80 \text{ m} = 115 \text{ m}$$

$$\text{Change of trim} = 12.00 - 11.20$$

$$= 0.80 \text{ m}$$

$$= 80 \text{ cm}$$

$$\text{Change of trim} = \frac{w \times d}{MCTC}$$

$$80 = \frac{w \times 115}{210}$$

$$w = 146 \text{ tonnes}$$

Change of trim after $= \dfrac{l}{L}$

$$= \dfrac{95}{200} \times 80$$

$$= 38 \text{ cm}$$

Change of trim forward $= 42$ cm

	F	A
Initial draft	11.20	12.00
Trim	$+0.42$	-0.38
Final draft	11.62	$11.62 =$ mean draft

Note that the average draft is 11.60 m, i.e. the mean draft is not equal to the average draft.

In practice it is often necessary to load a particular draft aft or forward. In *Figure 8.7*, a weight w is loaded over the centre of flotation, causing the vessel to sink from water line $W_0 L_0$ to $W_1 L_1$. If the weight is then moved a distance d m aft, the vessel will trim to waterline $W_2 L_2$. Thus the change of draft can be expressed as

$\pm$ Change of draft aft $= \pm$ sinkage $\pm$ change of trim aft

$$\pm \text{Change of draft aft} = \pm \dfrac{w}{TPC} \pm \dfrac{l \times w \times d}{L \times MCTC}$$

Sign conventions tend to be rather awkward to apply in these examples and students are advised to check their work carefully to ensure that they have been consistent in applying signs.

We will adopt the convention that an increase in draft is $+ve$, loading is $+ve$ and change of trim by stern in $+ve$.

Example 8.4

A vessel is about to enter a river port over a bar where the maximum depth at highwater is 9.2 m. She must have a minimum clearance of 0.5 m and is at present at draft. Forward 8.40 m, Aft 9.00 m tank. How much water must be discharged from an afterpeak tank LCG 7 m forward of AP?

TPC, 25 tonne/cm; MCTC, 125 tonne m/cm; LCF, midships; length, 220 m.

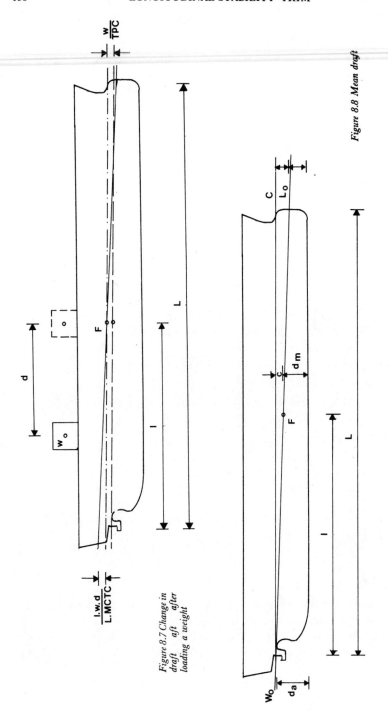

Figure 8.7 Change in draft aft after loading a weight

Figure 8.8 Mean draft

Maximum draft aft $= 8.70$ m
Present draft aft $= 9.00$ m

Change of draft aft $= 0.30$ m
$= 30$ cm
$d = 103$ m

Let w be ballast to discharge

$$\pm \text{Change in draft aft} = \pm \text{sinkage} \pm \text{change in trim aft}$$

$$-30 = -\frac{w}{TPC} - \frac{l}{L} \times \frac{w \times d}{MCTC}$$

$$30 = \frac{w}{25} + \frac{110 \times 103 \times w}{220 \times 125}$$

$$30 = 0.04w + 0.412w$$

$$30 = 0.452w$$

$$w = 66.4$$

Amount of ballast to discharge $= 66.4$ tonnes

$$\text{Change of trim} = \frac{w \times d}{MCTC}$$

$$= \frac{66.4 \times 103}{125} = 54.7 \text{ cm}$$

$$\text{Change of trim after} = \frac{l}{L} \text{Change of trim}$$

$$= \frac{110}{220} \times 54.7 = 27.4 \text{ cm}$$

Change of trim forward $= 27.4$ cm

$$\text{Rise} = \frac{w}{TPC}$$

$$= \frac{66.4}{25} = 2.7 \text{ cm}$$

	F	A
Initial draft	8.40 m	9.00 m
Rise	0.03 m	0.03 m
	8.37 m	8.97 m
Trim	+ 0.27 m	− 0.27 m
Final draft	8.64 m	8.70 m

Note that it is always necessary to check the final draft forward as it is possible that it could exceed the maximum permissable draft.

In the special case of keeping the draft aft, or forward, constant (*Figure 8.7*) we have

$$0 = \pm \text{sinkage} \pm \text{change of trim aft}$$

$$\text{sinkage} = \text{change of trim aft}$$

$$\frac{w}{TPC} = \frac{l}{L} \times \frac{w \times d}{MCTC}$$

From this equation, *d* represents the position to load a weight to keep the draft aft constant

$$d = \frac{L \times MCTC}{l \times TPC}$$

Note that the weight term does not appear in the equation. These are two limitations on the use of this equation:

1. The vessel could trim by the head until the forward draft is greater than the after draft.
2. The amount loaded could be greater than a moderate amount making the values of *MCTC* and *TPC* invalid.

In practice the position defined by *d* can be regarded as the centre of gravity of the weights loaded in compartments forward and aft of the position.

Example 8.5 (*Figure 8.9*)

A vessel floating at draft: Forward, 7.00 m; Aft, 8.00 m.

Distribute 600 tonnes of cargo between compartment 1 LCG 75 m forward to AP, and compartment 2 LCG 130 m forward of AP so as to maintain draft aft constant.

State the final draft forward.

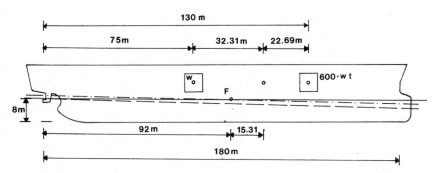

Figure 8.9 Loading to keep draft aft constant (Example 8.5)

TPC, 23 tonne/cm; MCTC, 180 tonne m/cm; LCF, 92 m forward of AP; length 180 m.

Distance of centre of gravity of weight from LCF

$$d = \frac{L \times MCTC}{l \times TPC}$$

$$= \frac{180 \times 180}{92 \times 23} \text{ m}$$

$$= 15.31 \text{ m forward of F}$$

LCF from AP = 92.00 m
g from AP = 107.31 m   107.31 m
Compartment 1 = 75.00 m Compartment 2   130.00 m
                ────────   ────────
                32.31 m aft   22.69 m forward
                ────────   ────────

Load w tonne aft

$$32.31w = 22.69 \ (600 - w)$$

$$55w = 22.69 \times 600$$

$$w = 248 \text{ tonne}$$

Load 248 tonne in Compartment 1
     352 tonne in Compartment 2

$$\text{Sinkage} = \frac{w}{TPC} = \frac{600}{23} \text{ cm} = 26 \text{ cm}$$

$$\text{Change of trim} = \frac{\text{moment changing trim}}{MCTC}$$

$$= \frac{w \times d}{MCTC}$$

$$= \frac{600 \times 15.31}{180} = 51 \text{ cm}$$

$$\text{Change of trim aft} = \frac{92}{180} \times 51 \text{ cm}$$

$$= 26 \text{ cm}$$

Change of trim forward = 25 cm

	F	A
Initial draft	7.00 m	8.00 m
	0.26 m	0.26 m
	7.26 m	8.26 m
Trim	+ 0.25 m	− 0.26 m
Final drafts	7.51 m	8.00 m

MEAN DRAFT

In Example 8.3 it was noted that the vessel trimmed to an even keel draft which was not the same as the average draft of the vessel before the weight was moved. Since a vessel trims about the centre of flotation parallel sinkage will always be the change in draft at the centre of flotation.

Thus, in determining the amount of weight to be loaded or discharged when moving from one trimmed draft to a different trimmed draft, it is necessary to determine the change in draft at the centre of flotation. In *Figure 8.8* (see page 198) the vessel length L is floating at a trimmed water line $W_0 L_0$ the centre of flotation is at F a distance 1 m from AP with the draft aft d_a the draft at $F d_m$ and draft forward d_f. $W_0 C$ is a line drawn parallel to the deck of the vessel.

Then $L_0 C$ the trim of the vessel
c = difference between d_a and d_m

Then using similar triangles

$$\frac{c}{1} = \frac{t}{L}$$

$$c = \frac{l}{L} \times t$$

$$d_m = d_a - \frac{l}{L} \times t$$

Note. In the unusual case of the vessel being trimmed by the head

$$dm = da + \frac{l}{L} \times t$$

Example 8.6 (*Figure 8.10*)

A vessel is floating at drafts: forward, 10.80 m; aft, 11.40 and is to complete loading at drafts: forward, 10.90 m; aft, 11.40 m; TPC, 28 tonne/cm; length, 138 m; LCF, 65 m from AP. Find the amount of cargo to load.

Initial mean draft

$$d_m = d_a - \frac{l}{L} \times t_I$$

$$= 11.40 - \frac{65}{138} \times 0.6$$

$$= 11.117 \text{ m}$$

Final mean draft

$$d_m = d_a - \frac{l}{L} \times t_F$$

$$= 11.4 - \frac{65}{138} \times 0.5$$

$$= 11.164 \text{ m}$$

Initial mean draft $= 11.117$ m

Sinkage $= 0.047$ m $= 4.7$ cm

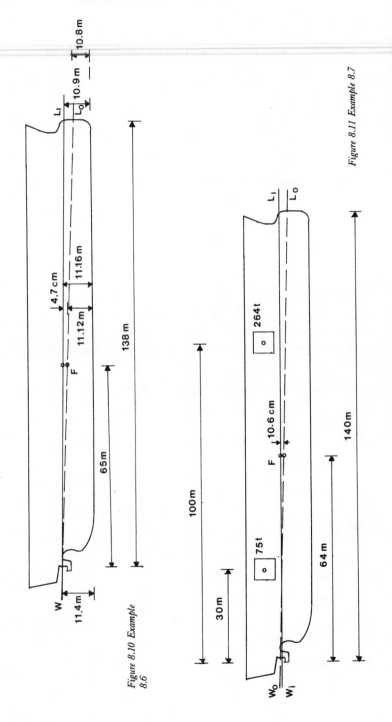

Figure 8.10 Example 8.6

Figure 8.11 Example 8.7

Cargo to load = sinkage × TPC

$$= 4.7 \times 28 \text{ tonne}$$

$$= 131.6 \text{ tonne}$$

If the calculation had been carried out between average drafts we would have

$$\text{Sinkage} = 11.15 - 11.10$$

$$= 0.05 \text{ m}$$

Cargo to load = 5 × 28

$$= 140.0 \text{ tonnes}$$

For large changes in draft the difference between displacements at the respective mean drafts should be used to determine the amount of cargo to be loaded or discharged.

Example 8.7 (*Figure 8.11*)

A vessel is floating at drafts: forward, 11.48 m; aft, 12.26 m. She is to complete loading at drafts: forward, 11.90 m; aft, 12.10 m.

Space is available in No 5 hold   30 m forward of AP
                      and No 2 hold 100 m forward of AP.

MCTC, 120 tonne m/cm; TPC, 32 tonne/cm
LCF, 64 m forward of AP; length, 140 m
Distribute the cargo to be loaded to give the required final drafts.

Initial mean draft

$$d_m = d_a - \frac{l}{L} \times t_1 \text{ m}$$

$$= 12.26 \text{ m} - \frac{64}{140} \times 0.78 \text{ m}$$

$$= 11.903 \text{ m}$$

Final mean draft

$$d_m = d_n - \frac{l}{L} t_m$$

$$= 12.10 \text{ m} - \frac{64}{140} \times 0.2 \text{ m}$$

$$= 12.009 \text{ m}$$

$$\text{Initial mean draft} = 11.903 \text{ m}$$

$$\text{Sinkage} = 0.106 \text{ m}$$

$$\text{Cargo to load} = \text{sinkage} \times \text{TPC}$$

$$= 10.6 \times 32 \text{ tonne}$$

$$= 339.2 \text{ tonne}$$

Let w be cargo to load in No 2 hold

$$\text{Change of trim} = \frac{\text{moment changing trim}}{\text{MCTC}}$$

Change of trim $= 78 \text{ cm} - 20 \text{ cm} = 58 \text{ cm}$ by head

$$-58 = \frac{34 \times (339.2 - w) - 36w}{120}$$

$$-6960 = 11\,532.8 - 34w - 36w$$

$$-6960 = 11\,532.8 - 70w$$

$$70w = 18\,492.8$$

$$w = 264.2 \text{ tonnes}$$

Load   264 tonne in No 2 hold
        75 tonne in No 5 hold

For completeness the following example is given to illustrate the calculation of *MCTC* from first principles.

Example 8.8

A vessel, length 200 m has the following $\frac{1}{2}$ ordinates of the 11 m waterplane commencing from aft are

Station	0	1	2	3	4	5	6	7	8	9	10
$\frac{1}{2}$ ord m	0	10.0	13.0	14.0	14.2	14.3	14.1	14.0	11.5	6.2	0.2

Under water volume to 11 m waterplane 64 300 m^3

KB 5.9 m KG 8.9 m

Vessel floating in salt water.
   Find the change of trim if a weight of 500 tonnes is shifted 60 m from aft to forward.

Find the $MCTC$.

Station	$\frac{1}{2}$ ord	SM	F (Area)	Lever	F (1st mom)	Lever	F (2nd mom)
0	0	1	0.0	0	0.0	0	0.0
1	10.0	4	40.0	1	40.0	1	40.0
2	13.0	2	26.0	2	52.0	2	104.0
3	14.0	4	56.0	3	168.0	3	504.0
4	14.2	2	28.4	4	113.6	4	454.4
5	14.2	4	56.8	5	284.0	5	1420.0
6	14.1	2	28.2	6	169.2	6	1015.2
7	14.0	4	52.0	7	364.0	7	2548.0
8	11.5	2	23.0	8	184.0	8	1472.0
9	6.2	4	24.8	9	223.2	9	2008.8
10	0.2	1	0.2	10	2.0	10	20.0
			335.4		1600.0		9586.4

$h \doteq 20$ m

Area

$$A = \frac{2}{3} h \times \sum F \text{ (area)}$$

$$= \frac{2}{3} \times 20 \times 335.4$$

$$= 4472.0 \text{ m}^2$$

$$I_{AP} = \frac{2}{3} h^3 \sum F \text{ (2nd moment)}$$

$$= \frac{2}{3} \times 20^3 \times 9586.4$$

$$= 51\,127\,467 \text{ m}^4$$

Centroid from AP

$$\bar{x} = \frac{h \times \sum F \text{ (1st moment)}}{\sum F \text{ (area)}}$$

$$= \frac{20 \times 1600.0}{335.4}$$

$$= 95.41 \text{ m}$$

$I_F = I_{AP} - A\bar{x}^2$

$$= 51\,127\,467 - 4472.0 \times 95.41^2$$

$$= 10\,418\,546 \text{ m}^4$$

$$BM_L = \frac{I_F}{\nabla}$$

$$KM_L = BM_L + KB$$

$$= \frac{10\,418\,546}{64\,300}$$

$$= 162.0\text{ m} + 5.9\text{ m}$$

$$= 162.0\text{ m}$$

$$= 167.9\text{ m}$$

$$GM_L = \quad 8.9\text{ m}$$

$$W = V\rho$$

$$GM_L = 159.0\text{ m}$$

$$= 64\,300 \times 1.025$$

$$= 65\,907.5\text{ tonne}$$

$$MCTC = \frac{W \times G_0 M_L}{100 \times L}$$

$$= \frac{65\,907.5 \times 159.0}{100 \times 200} \frac{\text{tonne m}}{\text{cm}}$$

$$= 524 \frac{\text{tonne m}}{\text{cm}}$$

$$\text{Change of trim} = \frac{L \times w \times d}{W \times GM_L}$$

$$= \frac{200 \times 500 \times 60}{65\,907.5 \times 159.0}\text{ m}$$

$$= 0.573\text{ m}$$

CHANGE IN TRIM DUE TO CHANGE IN DENSITY

A special case of change of trim occurs when a vessel moves between waters of different densities, if the centre of flotation and centre of buoyancy are not in the same vertical line.

In *Figure 8.12* a vessel displacement W is floating at waterline $W_0 L_0$ with centre of gravity at G and centre of buoyancy at Bo vertically beneath G. The centre of flotation is at F a horizontal distance d from Bo. The density of water is ρ_0, and there is parallel sinkage to waterline $W_1 L_1$, when water density is reduced to ρ_1.

A layer of buoyancy volume v has been added. Since the change in draft will be small, the centroid of this volume can be assumed to be at F the centre of flotation.

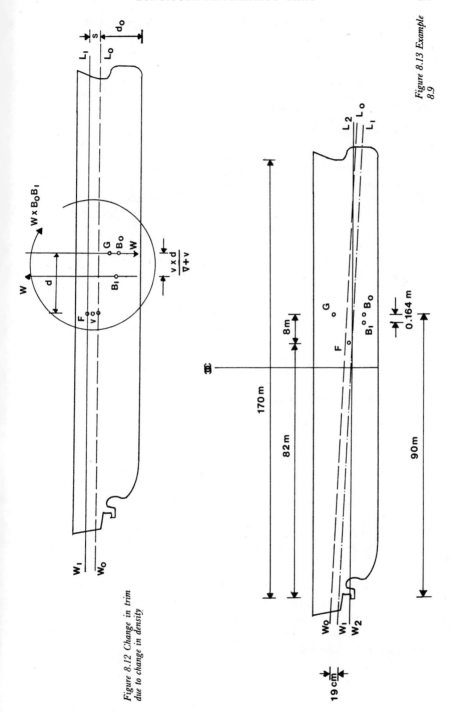

Figure 8.12 Change in trim due to change in density

Figure 8.13 Example 8.9

There will be a horizontal shift of the centre of buoyancy from B_0 to B_1. If the underwater volume of the vessel is ∇

$$B_0 B_1 = \frac{v \times d}{\nabla + v}$$

There will be a moment changing trim.

$$\text{Moment changing trim} = W \times B_0 B_1$$

$$\text{and Change of trim} = \frac{W \times B_0 B_1}{MCTC}$$

Sinkage can be found from FWA and DWA formula or from

$$\frac{do + s}{do} = \frac{\rho_0}{\rho_1}$$

$$do + s = do \times \frac{\rho_0}{\rho_1}$$

$$s = do \times \frac{\rho_0}{\rho_1} - do$$

$$s = do \times \left(\frac{\rho_0}{\rho_1} - 1 \right)$$

The increment of volume can be found from

$$v = \text{Sinkage} \times \text{area of waterplane}$$

$$v \simeq s \times \frac{TPC \times 100}{\rho_0}$$

or

$$(\nabla + v)\rho_1 = \nabla \rho_0$$

$$\nabla + v = \nabla \times \frac{\rho_0}{\rho_1}$$

$$v = \nabla \times \left(\frac{\rho_0}{\rho_1} - 1 \right)$$

Example 8.9 (*Figure 8.13*)

A vessel is floating at drafts: forward, 8.72 m, aft, 9.00 m in water density, 1.025 tonne/m³. She is to enter dock water density 1.004 tonne/m³. Find her drafts fore and aft in dock water, taking due account of the change of trim due to change of density.

$MCTC$, 162 tonne m/cm. TPC, 29.8 tonne/cm; LCF, 82 m forward of AP, LCB, 90 m forward of AP. Length, 170 m. Displacement, 27 000 tonnes.

Initial mean draft

$$d_{ml} = d_a - \frac{l}{L} \times t$$

$$= 9.00 - \frac{82}{170} \times 0.280$$

$$= 9.000 - 0.135$$

$$= 8.865 \text{ m}$$

Final mean draft

$$d_{mF} = d_{ml} \times \frac{\rho_1}{\rho_F}$$

$$= 8.865 \times \frac{1.025}{1.004} \text{ m}$$

$$= 9.050 \text{ m}$$

$$d_{ml} = 8.865 \text{ m}$$

$$s = 0.185$$

$$= 0.19 \text{ m}$$

$$\nabla = \frac{W}{\rho_0}$$

$$= \frac{27\,000}{1.025} \text{ m}^3$$

$$= 26\,341.5 \text{ m}^3$$

$$v = \nabla\left(\frac{\rho_1}{\rho_F} - 1\right)$$

$$= 26\,341.5 \times \left(\frac{1.025}{1.004} - 1\right)$$

$$= 551 \text{ m}^3$$

$$B_0B_1 = \frac{v \times d}{\nabla + v}$$

$$= \frac{551 \times 8}{26\,341.5 + 551} \text{ m}$$

$$= 0.164 \text{ m}$$

Since B has moved aft trim is by head

$$\text{Trim} = \frac{\text{moment changing trim}}{MCTC}$$

$$= \frac{W \times B_0B_1}{MCTC}$$

$$= \frac{27\,000 \times 0.164}{162} = 27.32 \text{ cm}$$

$$\text{Change of trim aft} = \frac{l}{L} \times \text{Change of trim}$$

$$= \frac{82}{170} \times 27.3 = 13.2 \text{ cm}$$

Change of trim forward $= 14.1$ cm

	F	A
Initial draft	8.72	9.00
Sinkage	0.19	0.19
	8.91	9.19
Trim	+0.14	−0.13
Final draft	9.05 m	9.06 m

CALCULATION OF TRIM WHEN HANDLING LARGE WEIGHTS

If a vessel is floating at a light trimmed draft and is to load to a load draft, the methods described above are unsatisfactory, as the value of $MCTC$ will change as will the position of the longitudinal centre of flotation. It is therefore necessary to use other methods.

The method most used in practice is to carry out the following steps.

1. Determine the position of the longitudinal centre of gravity in the initial condition.
2. Determine the position of the longitudinal centre of gravity in the final condition.
3. Using hydrostatic data find the position of the longitudinal centre of buoyancy, longitudinal centre of flotation, $MCTC$ and mean draft at the load displacement.
4. Find the trimming moment and hence change of trim by taking the moment composed of the horizontal distance between LCG and LCB and the displacement.

Determination of longitudinal centre of gravity

In *Figure 8.14* the vessel is floating at a trimmed water line W_0L_0 with displacement W. The centre of gravity G is vertically above the centre of buoyancy at B_0. Now if the vessel where trimmed by an external force about the centre of flotation F to float at the even keel waterline W_cL_c the centre of buoyancy would move to B_1. If the horizontal distance between B_0 and B_1 is x then the trimming moment is $W \times x$ and the trim t could be found from

$$t = \frac{W \times x}{MCTC}$$

$$x = \frac{t \times MCTC}{W}$$

The position of B_1 can be found from hydrostatic data as can the values of $MCTC$ and displacement.

Hence the position of the longitudinal centre of gravity G can be found.

Example 8.10 (*Figure 8.15*)

MV Nonesuch is floating at drafts: forward, 5.8 m; aft, 6.6 m. Determine the position of the longitudinal centre of gravity, length, 174 m.

$$\text{Average draft} = \frac{5.80 + 6.60}{2} = 6.2 \text{ m}$$

From hydrostatic data LCF 3.7 m forward of ⌷⌷

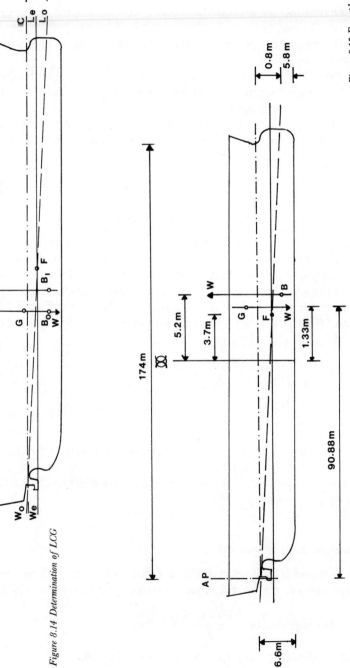

Figure 8.14 Determination of LCG

Figure 8.15 Example 8.10

$$\text{LCF from AP} = \frac{L}{2} + 3.7 \text{ m}$$

$$= 87.0 + 3.7$$

$$l = 90.7 \text{ m}$$

$$\text{Mean draft} = d_a - \frac{l}{L} \times t$$

$$= 6.6 - \frac{90.7}{174} \times 0.8$$

$$= 6.18 \text{ m}$$

Note. For most practical purposes the average draft will be sufficiently accurate.

At draft 6.18 m

$$MCTC = 356 \text{ tonne m/cm}$$

$$W = 21\,500 \text{ tonne}$$

$$LCB = 5.2 \text{ m forward of } ⌼$$

$$x = \frac{t \times MCTC}{W}$$

$$= \frac{80 \times 356}{21\,500} = 1.325 \text{ m}$$

Since vessel is trimmed by stern G is aft of even keel LCB

$$LCG ⌼ = 5.200 - 1.325$$

$$= 3.875 \text{ m}$$

$$\frac{L}{2} = 87.000 \text{ m}$$

$$\overline{}$$

$$LCG \text{ AP} = 90.875 \text{ m}$$

$$\overline{}$$

Example 8.11

MV Nonesuch is floating at drafts: forward, 2.90 m; aft, 4.7 m; length, 174 m.

She is to load cargo as follows

Compartment	Weight	LCG from AP
No 1	3800	146
No 2	4800	123
No 3	2200	100
No 4	5100	77
No 5	5300	51
ERDB	1000	22

Calculate the drafts of the vessel on completion.

Initial LCG

$$\text{Average draft} = \frac{2.90 + 4.7}{2} \text{ m}$$
$$= 3.80 \text{ m}$$

From hydrostatic data LCF, 5.3 m forward of ⌗

$$\text{LCF from AP} = \left(\frac{L}{2} + 5.3\right) \text{ m}$$

$$= 87 \text{ m} + 5.3 \text{ m}$$

$$= 92.3 \text{ m}$$

$$\text{Mean draft} = d_a - \frac{l}{L} \times t$$

$$= 4.7 - \frac{92.3}{174} \times 1.8$$

$$= 3.75 \text{ m}$$

From hydrostatic data

$$W = 12\,100 \text{ tonnes}$$

$$LCF = 5.3 \text{ m forward}$$

$$LCB = 6.0 \text{ m forward}$$

$$MCTC = 326 \text{ tonne m/cm}$$

$$x = \frac{t \times MCTC}{W}$$

$$= \frac{180 \times 326}{12\,100}$$

$$= 4.85 \text{ m}$$

Since vessel trimmed by stern G is aft of B

$$LCG \text{ from } AP = \frac{L}{2} + 6.0 - 4.85$$
$$= 87 + 6.0 - 4.85 = 88.15 \text{ m}$$

Compartment	Weight (tonne)	LCG from AP (m)	Moment (tonne m)
No 1	3 800	146	554 800
No 2	4 800	123	590 400
No 3	2 200	100	220 000
No 4	5 100	77	392 700
No 5	5 300	51	270 300
ERDB	1 000	22	22 000
Initial Cond	12 100	88.15	1 066 620
	34 300		3 116 820

$$LCG \text{ from } AP = \frac{\text{moment}}{\text{weight}}$$

$$= \frac{3 116 820}{34 300}$$

$$= 90.87 \text{ m}$$

$$= 3.87 \text{ m forward of } ⌼$$

From data at displacement 34 300 tonne

LCF, 1.6 m aft of $⌼$
LCB, 4.0 forward of $⌼$
$MCTC$ 432 tonne m/cm
mean draft 9.6 m

$$x = LCB - LCG$$

$$= 4.00 - 3.87 \text{ m aft of } B$$

$$= 0.13 \text{ m aft}$$

∴   Trim by stern

$$\text{Trim} = \frac{W \times X}{MCTC}$$

$$= \frac{34 300 \times 0.13}{432} = 10.3 \text{ cm}$$

$$LCF \text{ from } AP = \frac{L}{2} - LCF \text{ from } ⱷ$$

$$= 87 - 1.6$$

$$l = 85.4 \text{ m}$$

$$\text{Trim aft} = \frac{l}{L} \times t$$

$$= \frac{85.4}{174} \times 10.3$$

$$= 5 \text{ cm}$$

$$\text{Trim forward} = 5.3 \text{ cm}$$

	F	A
Mean draft	9.600	9.600
Trim	− 0.053	+ 0.050
	9.547	9.650

Final draft	Forward	9.55 m
	Aft	9.65 m

In practice the above example would be carried out on a proforma such as that shown in *Table 8.1*. This type of form is used to calculate fluid *GM* as well as the final drafts. There is no need to find the initial *LCG* as the work is always done from the light condition using, the lightship displacement, *LCG* and *KG*.

However, if there is insufficient data about the initial condition, it may be necessary to find the initial *LCG*. The choice of working about the *AP* or midships is a matter of personal choice. Working about the *AP* avoids negative numbers but results in rather large values.

Trimming tables

For small weights, work can be simplified by the provision of trimming tables. These tables ar graphs give the change in draft fore and aft as the result of loading some convenient unit of cargo say 100 tonnes, at any point along the length of the vessel. The information is normally given for load draft and ballast draft.

In *Figure 8.16* on a vessel length *L* m with centre of flotation as distance

Table 8.1 Proforma for determination of GM_f and draft

LOADING CALCULATION FORM

| | No. 5 | No. 4 | No. 3 | No. 2 | No. 1 |

▒ CARGO ▬ BALLAST ▨ FUEL OIL ⫿⫿⫿ FRESH W.

COLUMN No.	1	2	3 = 1x2	4	5 = 1x4	6
ITEM	WEIGHT (t)	ℓ G (m)	MOMENT (x10³ t-m)	K G (m)	MOMENT (x10³ t-m)	I P (t-m)
CARGO:						
No. 1 CARGO HOLD		-59.24				
No. 2 "		-36.04				
No. 3 " "		-12.90				
No. 4 " "		9.98				
No. 5 CARGO HOLD		35.92				
No. 1 TOP SIDE TK (P&S)		-57.58				
No. 2 " " (")		-35.60				
No. 3 " " (")		-12.90				
No. 4 " " (")		9.90				
No. 5 TOP SIDE TK (P&S)		36.30				
ON DECK CARGO						
CARGO TOTAL						
WATER BALLAST:						
No. 1 TOP SIDE TK (P&S)		-57.58				
No. 2 " " (")		-35.60				
No. 3 " " (")		-12.90				
No. 4 " " (")		9.90				
No. 5 TOP SIDE TK (P&S)		36.30				
No. 1 D.B.W.B.TK (P&S)		-58.87				
No. 2 " " (")		-35.56				
No. 3 " " (")		0.26				
No. 4 D.B.W.B.TK (C)		35.21				
FORE PEAK TANK		-75.11				
AFT PEAK TANK		77.21				
No. 3 C.H./W.B.TK		-12.90				
W.B. TOTAL						
FUEL OIL						
No. 3 D.B.F.O.TK (P&S)		0.29				
No. 4 D.B.F.O.TK (P)		36.02				
F.O. SETT. & SERV. TK	61	52.78	3.2	12.00	0.7	60
F.O. TOTAL						
DIESEL OIL:						
No. 4 D.B.D.O.TK (S)		36.02				
No. 5 D.B.D.O.TK (P)		55.43				
D.O. SETT. & SERV. TK	20	59.19	1.2	20.95	0.4	10
D.O. TOTAL						
FRESH WATER						
FRESH WATER TANK (P)		79.04				
FRESH WATER TANK (S)		77.16				
DIST. WATER TANK (P)		74.38				
F.W. TOTAL						
CONSTANT	158	48.13	7.6	12.15	1.9	-
DEADWEIGHT						
LIGHT WEIGHT	7,304	15.66	114.4	10.09	73.7	-
DISPLACEMENT						

DRAUGHT	CORRESPOND.	(m)		METACENTER ABOVE B.L. KM(m)
	FOR'D	(m)		C.OF GRAV. ABOVE B.L. KG(m)
	AFT	(m)		METACENTRIC HEIGHT CM(m)
	MEAN	(m)		EFFECT OF FREE SURFACE Gco(m)
TRIM		(m)		VERTICAL METACENT.HT. GoM(m)
C. OF GRAV. FROM ⊠ ⊠G		(m)		STILL MAX.BENDING HT. (t-m) (at Fr.)
C. OF BUOY. FROM ⊠ ⊠B		(m)		WATER MAX.SHEAR.F. (t) (at Fr.)
C. OF FLOAT. FROM ⊠ ⊠F		(m)		NOTES:-
MT.TO CHANGE TRIM 1cm MTC(t-m)				i) (-) SIGN SHOWS FORWARD FROM MIDSHIP
TONS PER 1cm IMMERSION TPC (t)				ii)HALF LOADED HOLD OR TANK IS ASTERISKES(*)

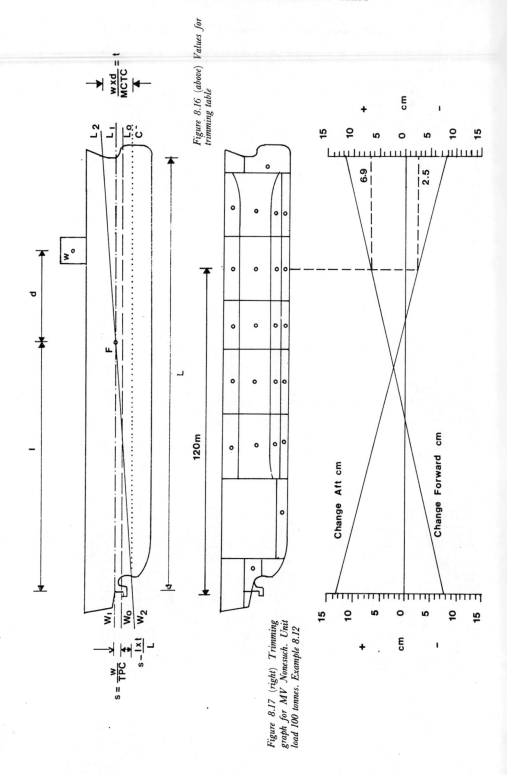

Figure 8.16 (above) Values for trimming table

Figure 8.17 (right) Trimming graph for MV Nonesuch. Unit load 100 tonnes. Example 8.12

1 m from the after perpendicular, the unit weight w is loaded a distance d m from the centre of flotation.

Then

$$\text{Change in draft aft} = \frac{w}{TPC} - \frac{l \times w \times d}{L \times MCTC}$$

For unit weight w/TPC is a constant K_0 and $(l \times w)/(L \times MCTC)$ is another constant K_1

$$\text{Change in draft aft} = K_0 + K_1 d$$

This is the equation of a straight line. Therefore by finding the values for the change in draft when the weight is loaded forward and again when the weight is loaded aft, a line can be drawn representing the change in draft aft after loading the unit weight at any point along the length of the vessel.

Similarly, for the same weight the change in draft forward will be

$$\text{Change in draft forward} = \frac{w}{TPC} + \frac{(L - l) \times w \times d}{L \times MCTC}$$

again $\dfrac{w}{TPC} = K_0$ and $\dfrac{(L - l) \times w}{L \times MCTC} = K_2$

The equation of the straight line is now

$$\text{Change in draft forward} = K_0 + K_2 d$$

The sign of the change in draft will depend upon the sign of d.

Example 8.11 *(Figure 8.17)*

Produce trimming table or graph for *MV Nonesuch* at displacement 35 000 tonnes assume length of vessel is 174 m.

Find the change of drafts if 400 tonne is loaded 120 m forward of *AP*.
From hydrostatic data.

Draft $= 9.75$ m
LCF   3.8 m forward of ⊛
TPC   39.3 tonne/cm

$l = 87 + 3.8 = 90.8$ m from *AP*

If weight is loaded at *FP* $d = -83.2$ m
If weight is loaded at *AP* $d = \quad 90.8$ m

$$K_0 = \frac{w}{TPC} = \frac{100}{39.3} = 2.54 \text{ cm}$$

$$K_1 = \frac{l \times w}{L \times MCTC} = \frac{90.8 \times 100}{174 \times 438} \frac{\text{cm}}{\text{m}}$$

$$= 0.119 \frac{\text{cm}}{\text{m}}$$

$$K_2 = \frac{(L - l)w}{L \times MCTC} = \frac{(174 - 90.8) \times 100}{174 \times 438}$$

$$= 0.109 \frac{\text{cm}}{\text{m}}$$

Change in draft aft

(i)   Loading at *FP*
      Change of draft aft $= K_0 + K_1 d$
      $$= 2.54 + 0.119 \times (-83.2)$$
      $$= -7.36 \text{ cm}$$

(ii) Loading at *AP*
      Change of draft aft $K_0 + K_1 d$
      $$= 2.54 + 0.119 \times 90.8$$
      $$= +13.35 \text{ cm}$$

Change in draft forward

(i)   Loading at *FP*
      Change of draft frd $= K_0 + K_2 d$
      $$= 2.54 + 0.109 \times 83.2$$
      $$= 11.61 \text{ cm}$$

(ii) Loading at *AP*
      Change of draft frd $= K_0 + K_2 d$
      $$= 2.54 + 0.109 \times 90.8$$
      $$= 12.44 \text{ cm}$$

Loading 400 tonnes 120 m forward of *AP*

Change forward from loading 100 tonnes $= +6.9$ cm
Total change $= 4 \times 6.9 = 27.6$ cm
Change aft from loading 100 tonne $= -2.5$ cm
Total change $= 4 \times 2.5$ cm $= 10$ cm

DETERMINATION OF DISPLACEMENT (DRAFT SURVEY)

In many cases when a vessel is loaded with bulk solid or liquid cargo, the only way of measuring the amount of cargo carried, is by finding the displacement of the ship.

On most ships displacement is presented as a scale of displacement against mean draft (see page 202). This displacement is calculated on the assumption that the vessel is on an even keel and is not hogged or sagged. In practice the ship will be trimmed and in the loaded condition slightly sagged or possibly hogged.

The problem is to determine displacement sufficiently accurately using available data such that if the ship loads w tonnes of cargo and

C_d = consumables on board on departure.
C_a = consumables on board on arrival.
K = constant for stores, crew, etc.
W_1 = light ship displacement.
W_d = measured displacement on departure.
W_a = measured displacement on arrival.

Then

$$W_d - (W_L + C_d + K) = W_a - (W_L + C_a + K) = w$$

In general W_d or W_a could be determined by measuring the underwater volume of the vessel ∇ and then

$$W = \nabla \times \rho_w$$

Unfortunately there is only rarely sufficient data available to enable this calculation to be carried out to the degree of accuracy required although this is an area where computers could make the calculation feasible in practice.

Some ships are provided with detailed forms and correction tables which apply to that ship only, in other cases shore-based operators require a standard procedure to be followed for all ships using a particular facility.

All the methods ultimately need to produce a value for draft d_t which after corrections for trim gives a value of W_d or W_a which is consistent with the above formula.

(*Note.* The value of w has to be consistent rather than accurate.)

A typical procedure would have the following steps:

1. Read density and find under keel clearance.
2. Read drafts port and starboard, forward
                port and starboard, midships
                port and starboard, aft

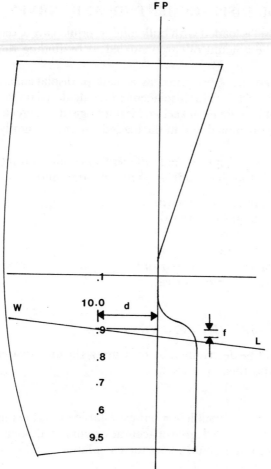

Figure 8.18 Correction of draft marks
to draft at FP

3. Take average of drafts port and starboard.
4. Correct average drafts forward and aft to give draft at fore perpendicular and after perpendicular. In *Figure 8.18* the draft marks are a distance d metres from the fore perpendicular then if the vessel is trimmed t metres by the stern and has length L the draft at the forepeak is

$$d_f = \text{average draft forward} - f$$

$$= \text{average draft forward} - \frac{t}{L} \times d$$

Similarly

$$d_a = \text{average draft aft} + \frac{t}{L} \times d$$

The actual sign of the correction will depend upon the trim and the position of the draft marks relative to the FP and the AP. In practice there may be correction tables.

5. In *Figure 8.19* the waterline is represented by WL indicating that the ship is sagged. (*Note.* The waterline is horizontal and the ship bent.) XX is a line drawn parallel to the bottom of the ship. If XX is positioned so that the volumes V_a, V_m and V_f between XX and waterline are made to be

$$V_m = V_a + V_f$$

then d_c is the draft on the displacement scale which will give the even keel displacement for the ship in the sagged condition

$$d_c = \frac{d_a + d_f}{2} + D$$

$$D = \text{constant} \times \left(d_m - \frac{d_a + d_f}{2} \right)$$

where d_m is the draft midships. If the waterline cuts XX a distance l m from midships

$$\text{Constant} = \frac{l}{\left(\dfrac{L}{2}\right)} = \frac{2 \times l}{L}$$

If l is expressed as a proportion of L investigation of ships forms leads to values of

$$\frac{2 \times l}{L} \text{ of between } \frac{2}{3} \text{ and } \frac{3}{4}$$

Substituting $\dfrac{2}{3}$ and $\dfrac{3}{4}$ for constant leads to values for d_c between

$$d_c = \frac{d_a + d_f + 4d_m}{6} \quad \left(\text{constant} = \frac{2}{3} \right)$$

$$d_c = \frac{d_a + d_f + 6d_m}{8} \quad \left(\text{constant} = \frac{3}{4} \right)$$

6. If the vessel is trimmed, a further correction is necessary. In *Figure 8.20* $t = d_a - d_f$. $W_1 L_1$ is a line drawn parallel to the waterline through d_c amidships. If the centre of flotation is a distance 1 m forward of midships and the ship has length L.

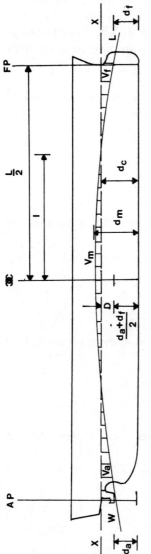

Figure 8.19 Correction for sagg (Hogg)

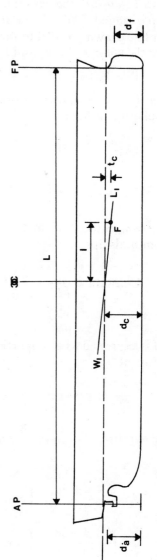

Figure 8.20 Correction for trim (I)

Then if the correction to d_c is t_c:

$$\frac{t_c}{l} = \frac{t}{L}$$

$$t_c = \frac{l}{L} \times t_m$$

The correction to the displacement found for d_c is

$$\text{correction } (1) = 100 \times t_c \times TPC \times \frac{\rho_{DW}}{\rho_{SW}}$$

The sign of the correction depends upon the direction of the trim and position of the *LCF* from midships. The sign can be obtained by inspection or from *Table 8.2*.

7. A further correction, generally referred to as Nemoto's correction, is sometimes applied. The need for a further correction arises because as the ship trims, the area of the water-plane passing through the centre of flotation must increase and hence the displacement increase as the volume beneath each successive trimmed waterline increases. (*Note.* This does not mean that trimming a ship increases displacement, it means that as successive waterlines through F are considered as trim increases, the volume beneath these waterlines increases.)

The formula is

$$\text{correction } (2) = \frac{t^2 \times 50}{Lpp} \times \frac{dM}{dZ}$$

where t is the trim in metres

Lpp is the length between perpendiculars

Table 8.2 Sign of correction for trim (1)

LCF / Trim	Forward of midships	Aft of midships
by head	-	+
by stern	+	-
	Correction	

$\dfrac{dM}{dZ}$ is the change in $MCTC$ over a range of drafts of

say $(d_c \pm t_c) + 0.5$ m and $(d_c \pm t_c) - 0.5$ m

The derivation of the formula is outside the scope of this book.

The correction is always positive. Unless trim is large the correction is small, i.e. for a typical ship displacing 30 000 tonnes trimmed 1 m by the stern the correction is:

$$\text{correction (2)} = \frac{t^2 \times 50}{Lpp} \times \frac{dM}{dZ}$$

$$= \frac{1 \times 50 \times 26.8}{174}$$

$$= +7.7 \text{ tonnes}$$

but if trim where 3 m

$$\text{correction (2)} = +69.3 \text{ tonnes}$$

A typical proforma for carrying at a draft survey is used in the example below (see *Table 8.3*).

Example 8.12

MV *Nonesuch* has the following draft when floating in water density 1.020 tonne m³.

Aft p	9.59 m	Fore p	8.07 m	Mid p	8.82 m
Aft s	9.65 m	Fore s	8.19 m	Mid s	9.20 m

Correction to forward draft	-0.02 m to FP
Correction to after draft	$+0.01$ m to AP
Under keel clearance	5 m
Length between perpendiculars	174 m

Heavy oil	894 tonnes
Diesel oil	102 tonnes
Lube oil	22 tonnes
Fresh water	152 tonnes
Ballast water	0
Constant	51 tonnes
Light ship	7304 tonnes

Find the amount of cargo on board. Use hydrostatic data from *Table 8.4*.

Table 8.3 Proforma for draft survey

M.V. Nonesuch   Draft Survey Form

Density D		1.020			
Underkeel Clearance		5m			

Drafts as read

Aft Port	9.59	Fore Port	8.07	Mid Port	8.82
Aft Stbd	9.65	Fore Stbd	8.19	Mid Stbd	9.20
Aft mean	9.62	Fore mean	8.13	Mid mean	9.01
Corr to AP	+ .01	Corr to FP	- .02		
da : (a)	9.63	df : (f)	8.11	dm : (m)	9.01
(f)	8.11	(a)	9.63		
(a - f)	1.52	(a + f)	17.74		
		(a + f)/2 : (e)	8.87	(e)	8.87
				(m - e)	.14
				2/3 (m - e)	.09
				(e)	8.87

$[(m)>(e)+, (m)<(e)-]:(e)\pm(2/3)\times(m)-(e) :dc:(c)$ 8.96m

Displacement at (c) from hydrostatics :	W_C	31904
Wcx(D/1.025)	W_D	31748

Correction for trim 1

		$100\times(a-f)$:(t)	152 tonne	
		LCF from midships:(1)	0.47	
		Lpp :(L)	174	
TPC from hydrostatics	38.05	TPCx(D/1.025) :(T)	37.9	
		$(T)\times(t)\times(1) /(L) :(C1)$	15.5	

Correction for trim 2

From hydrostatics			
MCTC at ((c)+0.5m)	429	$50X(a-f)^2\times(Z)/(L):(C2)$	19.3
MCTC at ((c)-0.5m)	400		
dm/mz :(2)	29		

		W_D	31748
Sign from Table (8.2) $\pm$		C1	+ 16
$+$		C2	19
		W_T	31783

Deductions

Heavy Oil	894
Diesel Oil	102
Lube Oil	22
Fresh Water	152
Ballast Water	0
Constant	51
Light ship	7304
W_K	8525

$-W_K$	8525
Cargo	23258 tonne

Table 8.4 Hydrostatics for *MV Nonesuch*

Draft	Displacement	TPC	MTC	LCB	LCF	KB	BM
14.00	52418	42.1	517.9	1.35f	3.47a	7.31	11.76
13.50	50320	41.8	509.3	1.56f	3.62a	7.05	11.65
13.00	48235	41.6	500.5	1.79f	3.72a	6.77	11.54
12.50	46162	41.4	491.0	2.04f	3.79a	6.51	11.46
12.00	44103	41.1	482.8	2.30f	8.70a	6.24	11.38
11.50	42055	40.8	474.6	2.59f	3.53a	5.98	11.33
11.00	40022	40.5	465.4	2.89f	3.23a	5.70	11.28
10.50	38002	40.2	455.8	3.21f	2.82a	5.44	11.26
10.00	36002	39.9	443.6	3.52f	2.20a	5.17	11.25
9.50	34019	39.5	429.4	3.83f	1.44a	4.91	11.29
9.00	32060	39.0	413.3	4.12f	0.53a	4.64	11.33
8.50	30122	38.6	400.5	4.40f	0.24f	4.39	11.43
8.00	28206	38.1	386.5	4.65f	1.18f	4.12	11.55
7.50	26309	37.7	374.3	4.87f	2.00f	3.87	11.75
7.00	24430	37.4	366.7	5.08f	2.60f	3.60	12.00
6.50	22565	37.1	358.9	5.26f	3.19f	3.35	12.35
6.50	22565	37.1	358.9	5.26f	3.19f	3.35	12.35
6.00	20714	36.8	351.7	5.43f	3.71f	3.08	12.77
5.50	18877	36.6	345.6	5.58f	4.15f	2.84	13.36
5.00	17052	36.3	339.8	5.72f	4.54f	2.56	14.08
4.50	15240	36.1	334.5	5.84f	4.88f	2.32	15.06
4.00	13440	35.8	328.8	5.96f	5.18f	2.05	16.31
3.50	11653	35.6	323.1	6.07f	5.43f	1.81	18.04
3.00	9883	35.2	316.3	6.18f	5.66f	1.53	20.36
2.50	8125	34.9	308.9	6.29f	5.84f	1.28	23.83
2.00	6397	34.4	298.9	6.41f	6.03f	1.02	28.87

QUESTIONS ON TRIM

1. The vessel to which the following data applies shifts 300 tonnes from a position 56 m forward of the *AP* to a position 105 m forward of the *AP*.

      Drafts   forward 8.92 m; aft 10.62 m
      MCTC   162 tonne m/cm

C of F   6 m aft of amidships
Length   124 m
Find the final drafts.

2.   The vessel to which the following data applies discharges 150 tonnes of cargo from a position 45 m forward of midships

Drafts   forward 8.37 m; aft 8.29 m
TPC   16.8 cm
MCTC   98 tonne m/cm
C of F   3 m aft of amidships
Length   100 m
Find the final drafts.

3.   A vessel is floating at drafts: forward, 8.60 m; aft, 9.20 m.
400 tonnes of cargo is loaded 126 m forward of *AP*.
250 tonnes is discharged 91 m forward of *AP*.
   Find the final drafts

MCTC   280 tonne m/cm
TPC   35 tonne cm
LCF   96 m forward of *AP*
Length   202 m

4.   A vessel is floating at drafts: forward, 9.87 m; aft, 10.50 m. She is to complete loading at an even keel draft of 10.40 m. Distribute the cargo to be loaded between spaces *LCG* 29 m forward of the *AP* and 126 m forward of the *AP*.

MCTC   180 tonne m/cm
TPC   25 tonne/cm
LCF   80 m forward of *AP*
Length   150 m

5.   A vessel is floating at drafts: forward, 6.00 m; aft, 6.70 m. The draft aft is to remain at 6.70 m. Distribute 1250 between spaces 38 m forward of *AP* and 118 m forward of *AP*.

MCTC   150 toone m/cm
TPC   25 tonne/cm
LCF   86 m forward of *AP*
Length   176 m

6.   A vessel is about to enter a river port over a bar where the maximum draft of water is 9.2 m. She must have a clearance of 0.5 m and is at present at drafts: forward, 8.40 m; aft, 9.00 m.

How much water must be discharged from the after peak tank
LCG 3 m forward of *AP* in order to achieve a safe draft aft.

MCTC	125 tonne m/cm
TPC	25 tonne/cm
LCF	106 m forward of *AP*
Length	212 m

7. A box shaped vessel length 60 m, breadth 10 m, depth 6 m is
    floating at an even keel draft of 4 m in salt water with a *KG* of 3 m.
    What will be the drafts of the vessel if a weight of 20 tonnes,
    already on board, is moved a distance 50 m along the vessel from
    forward to aft.

8. A vessel about to enter port is at drafts: forward, 11.20 m; aft,
    12.00 m. If the vessel is to enter port with an even keel draft, how
    much ballast must be transferred from a tank *LCG* 40 m forward of
    AP to a tank *LCG* 135 m forward of *AP*.
    What is the even keel draft?

MCTC	210 tonne m/cm
LCF	95 m forward of *AP*
Length	200 m

9. A vessel is floating at drafts: forward, 12.00 m; aft, 12.40 m.
    She loads cargo as follows

Weight	*LCG* from *AP*
Tonne	*AP* m
400	135
600	55

    She discharges

Weight	*LCG* from
Tonne	*AP* m
500	60 m

    Find the drafts on completion

MCTC	700 tonne m/cm
LCF	83 m forward of *AP*
TPC	46 tonne/cm
Length	170 m

10. A wall sided double bottom tank depth 1.5 m has half ordinates of
    breadth as follows at 4 m intervals commencing at the after
    bulkheads.

Station	0	1	2	3	4	5	6
½ ord m	6	6	5.9	5.7	5.4	4.9	4.3 m

The after bulkhead is 65 m forward of the centre of flotation.

Calculate the change of trim if the tank is filled with fuel oil density 0.95, *MCTC* 200 tonne metres/cm.

11. A vessel floating at drafts: forward, 9.48 m; aft, 10.40 m in water density 1.002 tonne/m³ and is to be on an even keel draft when in salt water.

How much ballast would you transfer from a tank *LCG*, 10 m forward of *AP* to a tank *LCG* 60 m forward of *AP*. What is the final draft?

MCTC	130 tonne m/cm
LCF	70 m forward of *AP*
LCB	65 m forward of *AP*
TPC	25 tonne/cm
Length	138 m
Displacement	22 000 tonnes.

12. *MV Nonesuch* is floating at drafts: forward, 4.20 m; aft, 5.98 m; length, 174 m in *SW*.

Cargo and ballast are handled as follows

Cargo loaded

Compartment	Weight (tonne)	LCG in from AP
No 1 hold	8100	146 m
No 2 hold	6500	100 m
No 5 hold	6900	51 m

Ballast discharged

Fore peak	890	162 m
No 1 *TSWT*	600	145 m
No 1 DB	800	146 m
No 2 DB	750	123 m
No 3 hold	5100	100 m
No 4 DB	400	52 m

Find the drafts of *MV Nonesuch* on completion.

TRIM (ANSWERS)

1. Final drafts   Forward, 9.42 m   Aft, 10.21 m

2. Final drafts   Forward, 7.89 m   Aft, 8.55 m

3. Final drafts   Forward, 8.88 m   Aft, 9.02 m

4. Sinkage 23.6 cm
    Load   163 tonnes,   29 m forward of *AP*
                          527 tonnes, 126 m forward of *AP*

5. Load   308 tonnes,   38 m forward of *AP*
                          942 tonnes, 118 m forward of *AP*

6. Discharge 66.4 tonnes
    Final draft   Forward, 8.64 m   Aft, 8.70 m

7. Final drafts   Forward, 3.835 m   Aft, 4.165 m

8. Even keel draft, 11.62 m
    Transfer, 146 tonnes

9. Final draft   Forward, 12.22 m   Aft, 12.40 m

10. Centroid of compartment 76.4 m forward of *LCF*
    Mass of oil   377.3 tonnes
    Change of trim   144.1 cm

11. Transfer 190 tonnes
    Final draft   Forward, 9.71 m   Aft, 9.71 m

12. Initial *LCG* 89.2 m forward of *AP*
    Final *LCG* 91.1 m forward of *AP*
    Final draft   Forward, 8.49 m   Aft, 8.67 m

9 Dry docking and grounding

This chapter deals with the stability and trim conditions which occur when the upthrust acting on the vessel is transferred from buoyancy to keel blocks or the ground.

OBJECTIVES

1. Description of the stability effects of the transfer of upthrust from buoyancy to the ground.
2. Assessment of stability during dry docking.
3. Assessment of stability at critical instant.
4. Assessment of stability as the tide falls after grounding at a single point.

DRY DOCKING

When a ship touches bottom and the water level is lowered, there is a steady transfer of upthrust from the centre of buoyancy to the point of contact with the ground. In *Figure 9.1* the ship has displacement W. Then if the upthrust through the keel (K) from the keel blocks in P, the upthrust through the centre of buoyancy will be $W - P$.

The resultant of these two forces will have magnitude W and will act at some point between B and K. The centre of gravity will remain at G and the displacement W will continue to act down through G. As the force P increases it can be seen that the resultant of the forces P and $W-P$ will get closer and closer to K. Eventually the resultant upthrust W will act in such a way that the ship will become unstable.

The danger in dry docking is that the ship may become unstable before additional support can be given using bilge blocks or possibly side shores.

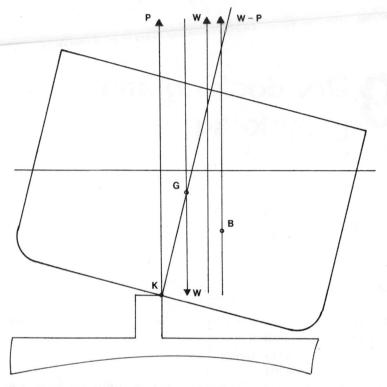

Figure 9.1 Transfer of upthrust from buoyancy to keel blocks

In the case of grounding the possibility of capsize is probably rather less as there will be larger areas of contact and the ship will only rarely be left high and dry.

ASSESSMENT OF REDUCTION IN KM (*Figure 9.2*)

A vessel displacement W is resting on keel blocks and is heeled to a small angle θ. The upthrust at the keel blocks is P the upthrust through the centre of buoyancy is $W-P$. The vector $W-P$ intersects the centreline at M_0.

The resultant of vector P and $W-P$ intersects the centreline at M_1 and has magnitude W.

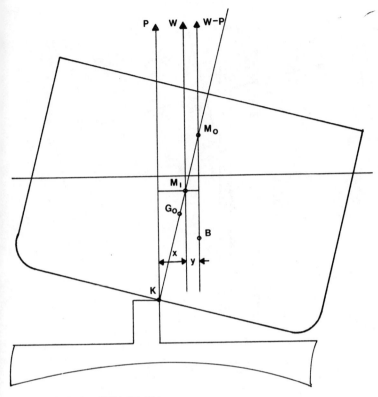

Figure 9.2 Reduction of KM (M_0M_1)

If the vector $W–P$ is a distance y from M_1 then

$$(W–P)y = Px$$
$$X = KM_1 \sin \theta$$
$$y = M_0M_1 \sin \theta$$
$$(W–P) \times M_0M_1 \times \sin \theta = P \times KM_1 \times \sin \theta$$
$$W \times M_0M_1 – P \times M_0M_1 = P \times KM_1$$
$$W \times M_0M_1 = P \times KM_1 + P \times M_0M_1$$
$$M_0M_1 = \frac{P \times KM_0}{W}$$

If the value of P is small, it can be assumed that the force $W–P$ intersects the centreline at the upright metacentre. If P is large the position of M_0 will be the position of the metacentre at displacement $W–P$.

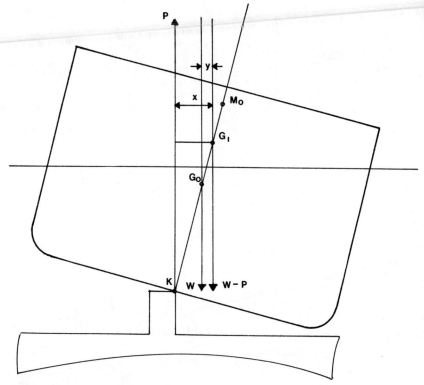

Figure 9.3 Effective rise of G (G_0G_1)

ASSESSMENT OF RISE IN EFFECTIVE KG (*Figure 9.3*)

In this case the resultant of the upthrust P acting through the keel and displacement W acting through G_0 is a force $W–P$ acting through G_1 above G_0.

If the vector P is a horizontal distance X from G_1 and the vector W a horizontal distance y from the vector through G_1

$$Wy = Px$$

$$X = KG_1 \sin \theta$$

$$y = G_0G_1 \sin \theta$$

$$W \times G_0G_1 \times \sin = P \times KG_1 \times \sin$$

$$W \times G_0G_1 = P \times KG_1$$

$$W \times G_0G_1 = P \times (KG_0 + G_0G_1)$$

$$W \times G_0G_1 = P \times KG_0 + P \times G_0G_1$$

$$(W{-}P)G_0G_1 = P \times KG_0$$

$$G_0G_1 = \frac{P \times KG_0}{W{-}P}$$

This formula can be interpreted as the rise in G resulting from the removal of a weight P from KG zero. Unlike the first formula this formula makes no assumptions about the point of intersection of the vertical through B_0 with the centreline is valid for all values of P.

Either of these formula can be used to find an approximate value for GM as water level falls. The values calculated will be sufficiently accurate for the early stages of docking and grounding. If an assessment of initial stability is necessary over a large range of drafts, the following methods may be used.

Example 9.1 (*Figure 9.4*)

MV Nonesuch is to be dry docked. She has mean draft of 5.00 m and KG 11 m.

Produce curves to give GM at drafts between 5 m and 2 m and from the curves estimate the draft at which the vessel becomes unstable.

From hydrostatic *Table 8.4*:

Draft (m)	Displacement 5 m draft	Displacement (W–P) (tonnes)	P (tonnes)	G_0G_1 $\left(\dfrac{P \times KG_0}{W{-}P}\right)$ (m)	KG_0	KG_1	KM_0	G_1M_0
5.0	17 052	17 052	0	0	11.00	11.00	14.08	3.08
4.5	17 052	15 240	1 812	1.31	11.00	12.31	15.06	2.75
4.0	17 052	13 440	3 612	2.96	11.00	13.99	16.31	2.32
3.5	17 052	11 653	5 399	5.10	11.00	16.10	18.04	1.94
3.0	17 052	9 883	7 169	7.98	11.00	18.98	20.36	1.38
2.5	17 052	8 125	8 927	12.09	11.00	23.09	23.83	0.74
2.0	17 052	6 397	10 655	18.33	11.00	29.33	28.87	−0.46

Draft at which *MV Nonesuch* becomes unstable, 2.18 m.

Note that from the above table the value of P is in general

$$P = \text{initial displacement} - \text{displacement during docking}$$

When a ship is being docked it is necessary to be sure that she will be

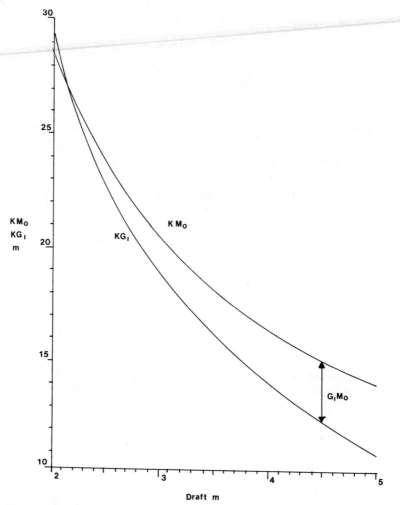

Figure 9.4 Assessment of effective GM at all drafts

stable until she touches the blocks fore and aft when additional support given by side shores and bilge blocks makes the vessel secure. In most cases the ship will enter dock trimmed by the stern. The ship will touch the blocks at or near the after perpendicular.

As the water level is lowered P will steadily increase and stability be reduced. The most critical moment will be just as the ship touches blocks fore and aft. This moment is called the critical instant.

P at the critical instant can be assessed as follows (see *Figure 9.5*).

The ship is floating at waterline W_0L_0 trimmed t cm by the stern with

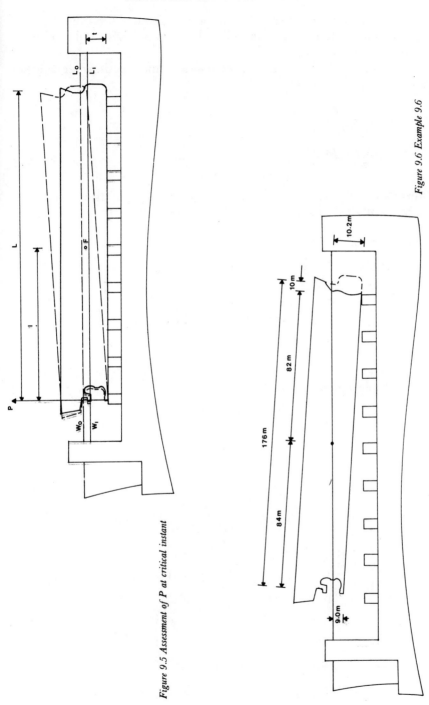

Figure 9.5 Assessment of P at critical instant

Figure 9.6 Example 9.6

displacement W. The centre of flotation is a distance 1 m from the after perpendicular. As the water level falls the ship will trim and P increase until she is at even keel.

The moment $P \times 1$ can be regarded as a moment causing a change of trim t cm

$$t \text{ cm} = \frac{P \times l}{MCTC}$$

$$P = \frac{t \times MCTC}{l}$$

This value of P at the critical instant can be used to assess the loss in stability at the moment the vessel takes the keel blocks fore and aft.

Example 9.2

A vessel about to dry dock is in the following condition.

Draft	forward, 6.10 m; aft, 6.70 m
KM_0	7.20 m; KG_0, 6.8 m
MCTC	155 tonne m/cm
TPC	22 tonne/cm
LCF	80 m forward of AP
Length	180 m
Displacement	11 000 tonnes

Find   (a) The GM of the vessel at the critical instant.
          (b) The righting moment at $1°$ heel.
          (c) The drafts fore and aft at the critical instant.

$$P = \frac{t \times MCTC}{l}$$

$$= \frac{60 \times 155}{80} = 116.3 \text{ tonne}$$

$$G_0G_1 = \frac{P \times KG_0}{W-P} \qquad\qquad M_0M_1 = \frac{P \times KM_0}{W}$$

$$= \frac{116.3 \times 6.80}{11\,000 - 116.3} \qquad\qquad = \frac{116.3 \times 7.20}{11\,000}$$

$$= 0.0727 \text{ m} \qquad\qquad\qquad\qquad = 0.0761 \text{ m}$$

$$G_0M_0 = 0.6000 \text{ m} \qquad\qquad\qquad G_0M_0 = 0.6000 \text{ m}$$

$$G_1M_0 = 0.5273 \text{ m} \qquad\qquad\qquad G_0M_1 = 0.5239 \text{ m}$$

Righting moment $= (W-P)G_1M_0 \sin \theta$

$$= (11\,000 - 116.3) \times 0.5273 \times \sin 1°$$

$$= 100.16 \text{ tonne}$$

Righting moment $= W \times W \times G_0M_1 \sin \theta$

$$= 11\,000 \times 0.5239 \times \sin 1°$$

$$= 100.56 \text{ tonne}$$

Within the limits of reasonable rounding the righting moments are equal. The apparent difference in the measurement of stability given by G_1M_0 and G_0M_1 can be explained in terms of righting moment

$$\text{'Bodily rise'} = \frac{P}{TPC}$$

$$= \frac{116.3}{22} \text{ cm}$$

$$= 5.3 \text{ cm}$$

$$\text{Change in trim aft} = \frac{1}{L} \times t$$

$$= \frac{80}{180} \times 60 = 26.7 \text{ cm}$$

Change in trim forward $= 33.3$ cm

	Fm	Am
Initial draft	6.100	6.700
Rise	0.053	0.053
	6.047	6.647
Trim	+0.333	−0.267
Draft on blocks	6.380	6.380

Example 9.3

A vessel about to dry dock is in the following condition.

Draft	forward, 5.62 m; aft, 6.82 m
KM_0	7.90 m; KG_0 7.40 m

MCTC	104 tonne m/cm
LCF	62 m forward of *AP*
Length	118 m
Displacement	8400 tonnes

At the critical instant the *GM* is to be no less than 0.45 m.

How much ballast should be transferred from a double bottom tank K_g 0.5 m, 30 m forward of *AP*, to a double bottom tank Kg 0.5 m, 90 m forward of the *AP* to ensure that the vessel will be in a satisfactory condition.

$$M_0M_1 = G_0M_0 - G_0M_1$$

$$= (0.50 - 0.45) \text{ m}$$

$$= 0.05 \text{ m}$$

$$M_0M_1 = \frac{P \times KM_0}{W}$$

$$P = \frac{M_0M_1 \times W}{KM_0}$$

$$= \frac{0.05 \times 8400}{7.9}$$

$$= 53.2 \text{ tonnes}$$

Maximum permissible trim

$$P = \frac{t \times MCTC}{l}$$

$$t = \frac{P \times l}{MCTC}$$

$$= \frac{53.2 \times 62}{104} \text{ cm}$$

$$= 31.7 \text{ cm}$$

Max. trim	31.7 cm
Present trim	120.0 cm
Change in trim	88.3 cm

Let ballast to shift be w tonnes.

$$\text{Change of trim} = \frac{\text{moment changing trim}}{MCTC}$$

$$88.3 = \frac{60 \times w}{104}$$

$$W = \frac{88.3 \times 104}{60}$$

$$= 153 \text{ tonnes}$$

Shift 153 tonnes of ballast forward.

Change of trim forward $= 41.9$ cm

	F	A
Initial drafts	5.62 m	6.82 m
Change in trim	+ 0.42 m	− 1.46 m
Final draft	6.04 m	6.36 m

Example 9.4

A vessel is to be drydocked and is in the following condition

Drafts

Forward	7.92 m; aft, 9.30 m
KM_0	11.43 m; KG_0, 10.90 m
MCTC	400.5 tonne m/cm
TPC	28.1 tonne/cm
LCF	88.5 m forward of AP
Length	174 m
Displacement	28 200 tonnes

The depth of water in the dock is initially 10.00 m. Find the effective GM of the vessel after the water level has fallen by 1.2 m in dock.

What are the drafts of the vessel after the fall?

Depth of water	10.00 m
After draft	9.30 m
Clearance	0.70 m
Fall	1.20 m
Change in draft aft	0.50 m = 50 cm

If P is the upthrust

$$\pm \text{Change in draft} = \pm \text{bodily rise} \pm \text{change in trim aft}$$

$$50 = \pm \frac{P}{TPC} \pm \frac{l \times P \times l}{L \times MCTC}$$

$$-50 = -\frac{P}{38.1} - \frac{88.5^2 \times P}{174 \times 400.5}$$

$$50 = 0.2626P + 0.1123P$$

$$50 = 0.1396P$$

$$P = 358.2 \text{ tonnes}$$

$$G_0G_1 = \frac{P \times KG_0}{W-P}$$

$$= \frac{358.2 \times 10.90}{28\,200 - 358.2}$$

$$= 0.140 \text{ m}$$

$$G_0M_0 = 0.530 \text{ m}$$

$$\overline{}$$

$$G_1M_0 = 0.390 \text{ m}$$

$$\text{Bodily rise} = \frac{P}{TPC}$$

$$= \frac{358.2}{38.1} \text{ cm}$$

$$= 9.40 \text{ cm}$$

$$\text{Change in trim} = \frac{P \times l}{MCTC}$$

$$= \frac{358.2 \times 88.5}{400.5} \text{ cm}$$

$$= 79.15 \text{ cm}$$

$$\text{Change of trim aft} = \frac{l}{L} \times t$$

$$= \frac{88.5}{174} \times 79.2 = 40.6 \text{ cm}$$

$$\text{Change of trim frd} = 38.9$$

	F_m	A_m
Initial draft	7.920	9.300
Bodily rise	0.094	0.094
	7.826	9.206
Trim	+0.389	−0.406
Drafts after fall	8.215	8.800

Example 9.5 (*Figure 9.6; see page 241*)

A vessel has been damaged forward and is to be docked in the following condition taking into account effective of damage on hydrostatic data. Waterline intersects forward perpendicular at 10.20 m; Draft, aft 9.00 m.

Vessel touches the blocks 10 m aft of the forward perpendicular

KM_0	11.25 m; KG_0 10.6 m
$MCTC$	440 tonne m/cm
TPC	39.5 tonne/cm
LCF	84 m forward of AP
Length	176 m
Displacement	35 500 tonnes

Find the effective GM of the vessel when she takes the blocks fore and aft. What will be the draft on taking the blocks.

In this case l is from 10 m aft to FP to LCF.

$$l_b = (L - l_i) - 10$$

$$= (176 - 84) - 10 \text{ m}$$

$$= 82 \text{ m}$$

$$P = \frac{MCTC \times t}{l_b}$$

$$= \frac{440 \times 120}{82} = 643.9 \text{ tonnes}$$

$$M_0 M_1 = \frac{P \times KM}{W}$$

$$= \frac{644 \times 11.25}{35\,500}$$

$$= 0.204 \text{ m}$$

$$G_0 M_0 = 0.650 \text{ m}$$

$$G_0 M_1 = 0.446 \text{ m}$$

$$\text{Bodily rise} = \frac{P}{TPC}$$

$$= \frac{644}{39.5} \text{ cm} = 16.3 \text{ cm}$$

$$\text{Change of trim aft} = \frac{l}{L} \times t$$

$$= \frac{84}{176} \times 120 = 57.3 \text{ cm}$$

$$\text{Change of trim frd} = 62.7 \text{ cm}$$

	F_m	A_m
Initial draft	10.200	9.000
Rise	0.163	0.163
	10.037	8.837
Trim	−0.627	+0.573
Draft on blocks	9.410	9.410

GROUNDING

When a ship grounds at a single point it is possible to make an estimate of the upthrust at the grounding point and the changes in mean draft and trim if the fall in tide is known.

The method used is an extension of the relationship used to calculate change in draft aft in Chapter 8. In the case of grounding we know the change in draft at the point of contact, if we know the fall in tide. In *Figure 9.7* the vessel length L m grounds at a point C, x m forward of the centre of flotation. The tide falls a distance y cm and the waterline changes from $W_0 L_0$ to $W_1 L_1$. The change in draft at the point of contact is the same as the fall of tide y cm.

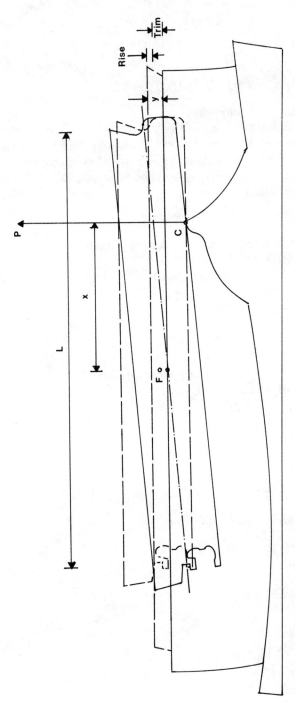

Figure 9.7 Grounding at a single point

Then

Change of draft at C = bodily rise + change of trim at C

$$y = \frac{P}{TPC} + \frac{x}{L} \times \frac{P \times x}{MCTC}$$

from which P can be found.

Once P is found it is possible to calculate the new drafts of the vessel by treating P as a weight discharged from C. Similarly the effect on stability can be calculated. The method must be regarded as approximate as the change in water level may be large and the value of TPC, $MCTC$ and X will change considerably.

Example 9.6

A vessel floating at drafts: forward, 8.70 m; aft, 9.40 m, grounds at a point 30 m aft of the forward perpendicular.

Estimate the drafts of the vessel and the GM after the tide has fallen by 70 cm.

$MCTC$	340 tonne m/cm
TPC	28 tonne/cm
KG_0	7.60 m, KM_0 8.40 m
Length	162 m
LCF	78 m forward of AP
Displacement	29 000 tonne

$$y = \frac{P}{TPC} + \frac{X \times P \times X}{L \times MCTC}$$

$$70 = \frac{P}{28} + \frac{54 \times 54 \times P}{162 \times 340}$$

$$70 = 0.0357P + 0.0529P$$

$$P = 789.7 \text{ tonnes} \simeq 790 \text{ tonne}$$

$$G_0 G_1 = \frac{P \times KG_0}{W - P}$$

$$= \frac{790 \times 7.80}{29\,000 - 790}$$

$$= 0.218 \text{ m}$$

$$G_0 M_0 = 0.700 \text{ m}$$

$$\overline{}$$

$$G_1 M_0 = 0.482 \text{ m}$$

$$\text{Bodily rise} = \frac{P}{TPC} = \frac{790}{28} \text{ cm}$$

$$= 28 \text{ cm}$$

$$\text{Change in trim} = \frac{P \times X}{MCTC}$$

$$= \frac{790 \times 54}{340} \text{ cm}$$

$$= 125.5 \text{ cm}$$

$$\text{Change in trim aft} = \frac{l}{L} \times t$$

$$= \frac{78}{162} \times 125.5 \text{ cm}$$

$$= 60 \text{ cm}$$

Change in trim frd = 65.5 cm

	Forward (m)	Aft (m)
Initial draft	8.70	9.40
Rise	− 0.28	− 0.28
	8.42	9.12
Trim	− 0.66	+ 0.60
Final draft	7.76	9.72

QUESTIONS ON DRY DOCKING AND GROUNDING

1. A vessel is to be dry docked and the reduction in GM is to be no more than 0.1 m before the critical instant.
   Find the maximum permissable trim

$MCTC$	146 tonne m/cm
LCF	70 m forward of AP
KG_0	8.40 m; KM_0, 9.0 m
Displacement	14 000 tonnes

2. A vessel is to be dry docked and is at draft: forward, 7.80 m; aft, 8.90 m. Find
    (a) *GM* at critical instant.
    (b) Righting moment at 1° heel at critical instant.
    (c) Drafts fore and aft at critical instant.
    (d) Upthrust acting on vessel when water level has fallen 20 cm after critical instant.

MCTC	172 tonne m/cm
TPC	27 tonne/cm
LCF	92 m forward of *AP*
Length	176 m
KG_0	7.50 m; KM_0, 8.4 m
Displacement	12 500 tonne

3. A vessel is to be dry docked and is floating at drafts: forward, 6.00 m; aft, 7.00 m.
    Given the following data is it safe to drydock the vessel?
    MCTC 460 tonne m/cm

KG_0	9.0 m; KM_0, 9.3 m
Length	300 m
LCF	153 m forward of *AP*
Displacement	12 500 tonne

4. A vessel about to be dry docked is floating at drafts: forward, 9.90 m; aft, 10.70 m. She has

KG_0	9.4 m; KM_0, 10.2 m
MCTC	360 tonne m/cm
TPC	34 tonne/cm
LCF	82 m forward of *AP*
Length	160 m
Displacement	23 500 tonnes

    Find
    (a) *GM* at critical instant
    (b) Drafts as critical instant
    (c) *GM* when draft is 10.00 m

5. A vessel is to be dry docked in a damaged condition. She is floating at drafts forward 12.60 m, aft 9.00 m. It is estimated that she will take the blocks 15 m aft of the fore perpendicular. Given the following data, is it safe to dry dock the vessel in this condition?

Estimated data

KG_0	8.9 m; KM_0 9.4 m
TPC	21.3 tonne/cm
$MCTC$	290 tonne m/cm
LCF	76 m forward of AP
Length to FP	165 m
Displacement	19 500 tonnes

Estimate the drafts at the perpendiculars on taking the blocks fore and aft.

6. A vessel length 130 m is floating at drafts forward 6.8 m; aft 7.6 m goes aground at a single point 20 m aft of the fore perpendicular. The tide falls by 1 m. Given the following data estimate the GM of the vessel and the drafts after the fall of tide.

KG_0	6.2 m; KM_0 7.0 m
$MCTC$	180 tonne m/cm
TPC	20 tonne/cm
LCF	64 m forward to AP
Displacement	18 000 tonne

DRY DOCKING AND GROUNDING (ANSWERS)

1. 75 cm by stern.

2. P at critical instant 170 tonne
   (a) $G_0M_1 = 0.797$ m; $G_1M_0 = 0.786$ m
   (b) Righting moment 171.5 tonne m
   (c) Drafts forward and aft 8.262 m
   (d) P after addition 20 cm fall $= (170 + 540)$ tonne
   $\qquad\qquad\qquad\qquad\qquad = 710$ tonne

3. $P = 301$ tonne
   $G_1M_0 = 0.206$ m safe but marginal.

4. At critical instant
   $P = 351$ tonnes $\qquad G_1M_0 = 0.656$ m
   $\qquad\qquad\qquad\qquad G_0M_1 = 0.648$ m
   Draft 10.28 m fore and aft
   At draft 10.00 m
   $P = 1303$ tonnes $\qquad G_1M_0 = 0.265$ m
   $\qquad\qquad\qquad\qquad G_0M_1 = 0.236$ m

5. $P = 1411$ tonnes
    $G_1 M_0 - 0.194$ m $\qquad G_0 M_1 - 0.180$ m
    Therefore unsafe
    Drafts 10.000 m $\qquad$ Forward and aft.

6. Drafts forward, 5.54 m; aft, 8.17 m
    $G_1 M_0$ 0.54 m; $G_0 M_1$, 0.52 m.

10 Bilging

In this chapter the effect on draft and stability of flooding will be considered. The work will be restricted to simple box shapes and is intended to given an indication of the principles involved rather than a full description of the process of determining floodable length and the production of flooding curves.

OBJECTIVES

1. To determine the draft and stability condition of a box shape after a symmetrical midship compartment is bilged.
2. To determine the permeability of a compartment containing cargo.
3. To determine the list of a box vessel when a midship side compartment is bilged.
4. To determine the trim of a box vessel when an end compartment is bilged.
5. Description of the use of flooding curves.

FLOODING AND STABILITY

When a compartment is laid open to the sea and flooded it is said to be bilged. There are two possible ways of viewing the flood water. Firstly the water can be regarded as an addition to the displacement of the vessel. This approach has several disadvantages such as the difficulty in assessing the amount of water in the compartment, the need to take into account the change in KG and the effects of the free surface of the added water.

The second approach is to consider the bilged compartment as having been removed from the vessel so that it no longer contributes to the buoyancy of the vessel. This approach has several advantages, i.e. the displacement of the vessel does not change nor does the position of the

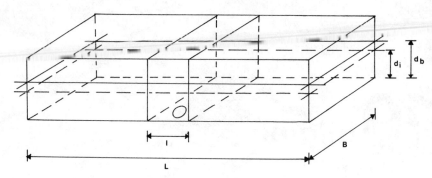

Figure 10.1 Bilging of midship compartment

centre of gravity, there is no need to consider free surface effects. The lost buoyancy, or constant displacement approach also lends to a very simple relationship between the condition of the vessel before and after bilging.

Intact underwater volume before bilging
= Intact underwater volume after bilging

In *Figure 10.1* a box vessel has length L, breadth B and is floating at draft d_i. Amidships centre line compartment length l, breadth b extending the full depth of the vessel is bilged. Then if the bilged draft is d_b,

Intact volume before bilging = Intact volume after bilging

$$LBd_i = LBd_b - lbd_b$$

from which d_b can be determined.

It is nearly always more convenient to work by subtracting the underwater volume of the bilged compartment from the total underwater dimensions of the vessel in the bilged condition. Since there are many different possibilities for the form of the bilged compartment even for a simple box shape, there is little profit in attempting to produce a general expression for d_b, the bilged draft.

The stability of the vessel will be affected because the position of the centre of buoyancy will move and the inertia of the waterplane will be altered as a result of the changes in area of the waterplane.

The general relationship that:

$$KM = KB + BM = KB + \frac{I}{\nabla}$$

holds good for the bilged condition. However, it is not possible to produce a general relationship for KM and each case must be examined individually, bearing in mid that the underwater volume ∇ remains constant, when compared with the intact condition. Some examples will amplify the possibilities.

Example 10.1

A box vessel: length, 110 m; breadth, 12 m; depth, 8 m is floating at draft 6 m. A midships compartment extending the full breadth of the vessel length 9 m is bilged. If the vessel has KG 4.8 m and is floating in salt water. Find

(a) the bilged draft;
(b) the GM of the vessel in the initial condition;
(c) the GM of the vessel in the bilged condition;
(d) the GM of the intact vessel at the bilged draft;
(e) the righting moment of the vessel at $1°$ heel in conditions (b), (c) and (d).

Intact volume before bilging = Intact volume after bilging

$$LBD_i = LBd_b - lBd_b$$
$$110 \times 12 \times 6 \text{ m}^3 = 110 \times 12 \times d_b - 9 \times 12 \times d_b$$
$$7920 \text{ m}^3 = 1320d_b - 108d_b$$
$$7920 \text{ m}^3 = 1212d_b$$
$$d_b = 6.535 \text{ m}$$

Displacement $= LBd_i \rho sw$

$$= 7920 \times 1.025 \text{ tonne}$$
$$= 8118 \text{ tonne}$$

KM intact

$$KM = \frac{d_b}{2} + \frac{B^2}{12d_b}$$
$$= \frac{6}{2} + \frac{12^2}{12 \times 6}$$
$$= 3 + 2 = 5 \text{ m}$$

$GM = 5 - 4.8 = 0.2$ m
Righting moment $= W \times GM \times \sin \theta$

$$= 7920 \times 0.2 \times \sin 1° \text{ tonne m}$$
$$= 27.64 \text{ tonne m}$$

KM bilged

$$KM = KB + \frac{I}{V}$$

$$= \frac{d_b}{2} + \frac{(L-l) \times B^2}{12(L-l) \times B \times d_b}$$

$$= \frac{d_b}{2} + \frac{B^2}{12d_b}$$

$$= \frac{6.535}{2} + \frac{144}{12 \times 6.535}$$

$$= 3.268 + 1.836$$

$$= 5.104 \text{ m}$$

$$KG = 4.800 \text{ m}$$
$$GM = 0.304 \text{ m}$$

Righting moment $= W \times GM \times \sin \theta$

$$= 7920 \times 0.304 \times \sin 1°$$

$$= 52.02 \text{ tonne m}$$

KM intact at bilged draft

$$KM = \frac{d_b}{2} + \frac{B^2}{12d_b}$$

$$= \frac{6.535}{2} + \frac{144}{12 \times 6.535}$$

$$= 5.104 \text{ m}$$

$$KG = 4.800 \text{ m}$$

$$GM = 0.304 \text{ m}$$

Intact displacement at $6.535 = L \times B \times d_b \times \rho$

$$= 110 \times 12 \times 6.535 \times 1.025$$

$$= 8841.9 \text{ tonnes}$$

Righting moment $= W \times GM \times \sin 1°$

$$= 8841.9 \times 0.304 \times \sin 1°$$
$$= 49.91 \text{ tonne m}$$

In this example the bilged GM is greater than the initial GM.

This is not a general result, a change in the dimensions of the vessel or the intact draft could have produced a reduction in GM. In the case of a full breadth compartment being bilged, the bilged GM of a box vessel is the same as GM for the intact vessel at the bilged draft.

However, since the bilged vessel has less displacement than the intact vessel, the righting moment is reduced and the vessel is less stable than it would have been had it been intact. Also the freeboard is reduced with the consequent effect on the GZ curve as described in Chapter 6. If the compartment does not extend the full breadth and/or depth of the vessel the situation is modified.

Example 10.2 (*Figure 10.2*)

A box vessel; length, 80 m; breadth, 6 m; depth, 4 m; is floating at draft 2 m, KG, 2.2 m. A midship centraling compartment length, 10 m; breadth, 4 m extending the full depth of the vessel is bilged.

Find the bilged draft and the GM in the intact and bilged condition.

Intact volume before bilging = intact volume after bilging

$$LBd_i = LBd_b - lbd_b$$

$$80 \times 6 \times 2 = 80 \times 6 \times d_b - 10 \times 4 \times d_b$$

$$960 = 480d_b - 40d_b$$

$$d_b = \frac{960}{440}\ \mathrm{m} = 2.182\ \mathrm{m}$$

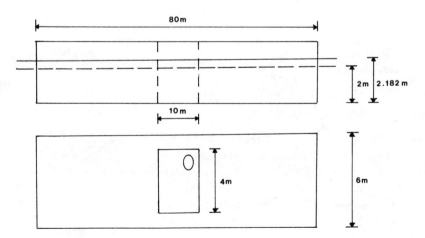

Figure 10.2 Example 10.2

KM intact

$$KM = \frac{d_i}{2} + \frac{B^2}{12d_i}$$

$$= 1 + \frac{36}{12 \times 2} = 2.5 \text{ m}$$

$$GM_1 = KM_1 - KG$$
$$= 2.5 - 2.2$$
$$= 0.3 \text{ m}$$

KM bilged

$$KM = KB + \frac{I}{V}$$

$$KB = \frac{d_b}{2} = \frac{2.182}{2} = 1.091 \text{ m}$$

$$I = \frac{LB^3}{12} - \frac{lb^3}{12}$$

$$= \frac{80 \times 6^3 - 10 \times 4^3}{12}$$

$$= \frac{17\,280 \times 640}{12} = 1386.7 \text{ m}^4$$

$$KM = 1.091 + \frac{1386.7}{960}$$

$$= 1.091 + 1.444 = 2.535 \text{ m}$$

$$GM_1 = KM - KG$$
$$= 2.535 - 2.200$$
$$= 0.335 \text{ m}$$

Example 10.3 (*Figure 10.3*)

A box vessel length 140 m; breadth, 20 m; depth, 14 m; is floating at draft 5.8 m; KG 8.00 m. A midship compartment length 24 m, breadth 7 m, has a watertight flat at a height 6 m above the keel. Find the bilged draft and the GM in the intact and bilged condition.

In this example there are two possibilities:

1. The watertight flat will be immersed;
2. The flat is not immersed.

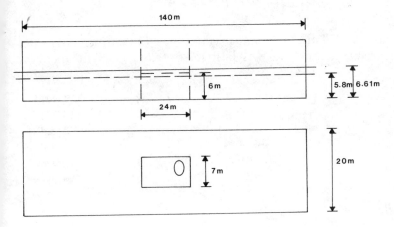

Figure 10.3 Example 10.3

If the flat is not immersed the example is similar to Example 10.2. Assuming the flat is immersed. Then if the flat is a distance L m above the keel.

Intact volume before bilging = Intact volume after bilging

$$LBd_i = LBd_b - lbh$$
$$140 \times 20 \times 5.8 = 140 \times 20 \times d_b - 24 \times 7 \times 6$$
$$16\,240 = 2800d_b - 1008$$
$$d_b = \frac{17\,248}{2800}$$
$$= 6.16 \text{ m}$$

KM intact

$$KM = \frac{d_i}{2} + \frac{B^2}{12d_i}$$
$$= \frac{5.8}{2} + \frac{20^2}{12 \times 5.8}$$
$$= 2.9 + 5.747 \text{ m} = 8.647 \text{ m}$$
$$GM = KM - KG$$
$$= 8.647 - 8.000$$
$$= 0.647 \text{ m}$$

KM bilged

In this case the KB must take into account the fact that the underwater shape is not symmetrical. KB must be found by taking moments about the keel.

	Volume	Kb	Moment
$L \times B \times d_b$	17 248	3.08	53 123.8
$l \times b \times h$	$-1\,008$	3.00	$-3\,024.0$
	16 240		50 099.8

$$KB = \frac{\text{moment}}{\text{volume}}$$

$$= \frac{50\,099.8}{16\,240} = 3.085 \text{ m}$$

In this case the waterplane is intact

$$BM = \frac{I}{V}$$

$$= \frac{LB^3}{12LBd_i}$$

$$= \frac{B^2}{12d_i}$$

$$= \frac{20^2}{12 \times 5.8} = 5.747 \text{ m}$$

$$GM_1 = KB + BM - KG$$

$$= 3.085 + 5.747 - 8.000$$

$$= 0.832 \text{ m}$$

PERMEABILITY

If a compartment contains cargo machinery, etc then the flood water cannot occupy the whole of the compartment. The proportion of the compartment which can be occupied by water is called the permeability of the compartment.

$$\text{Permeability } (\mu) = \frac{\text{Volume of compartment occupied by water}}{\text{Volume of compartment}}$$

In most practical cases permeability must be estimated. For example a typical 'empty' compartment μ will be about 95%. Permeability will be high for cargos such as cars and low for many bulk cargoes.

Permeability can be calculated in special cases where both the density and stowage factor are known.

If cargo has density ρ tonne/m³ and stowage factor SF m³/tonne. Then

1 tonne of cargo stowed solid occupies $\dfrac{1}{\rho} \dfrac{\text{m}^3}{\text{tonne}}$

1 tonne of cargo as stowed occupies SF m³/tonne

Space available for water $= \left(SF - \dfrac{1}{\rho} \right)$ m³/tonne

$$\text{Permeability} = \frac{\left(SF - \dfrac{1}{\rho} \right)}{SF}$$

or as a percentage

$$\mu = 100 \frac{\left(SF - \dfrac{1}{\rho} \right)}{SF}$$

Example 10.4

A box vessel length, 100 m; breadth, 9 m; depth, 6 m; is floating at craft 5 m. A full breadth midship compartment length 20 m contains cargo stowage factor 1.2 m³/tonne density 2 tonne/m³ find the bilged draft.

$$\mu = \frac{SF - \dfrac{1}{\rho}}{SF} = \frac{1.2 - \frac{1}{2}}{1.2}$$

$$= \frac{0.7}{1.2} = 0.583$$

Intact volume before bilging = Intact volume after bilging

$$LBd_i = LBd_b - lBd_b\, \mu$$

$$100 \times 9 \times 5 \text{ m}^2 = 100 \times 9 \times d_b - 20 \times 9 \times 0.583 \times d_B$$

$$4500 \text{ m}^3 = 900 \times d_B - 104.9 d_B$$

$$d_b = 5.65 \text{ m}$$

If there is cargo in a compartment the change in stability can best be dealt with by reducing the second moment of the bilged area in proportion to the permeability, i e. in Example 10.4.

$$KM = \frac{d_b}{2} + \frac{I}{V}$$

$$I = \frac{lB^3}{12} - \frac{lB^3}{12}\mu$$

$$= \frac{100 \times 9^3 - 20 \times 9^3 \times 0.583}{12}$$

$$= 5366 \text{ m}^4$$

$$KM = 2.83 \text{ m} + \frac{5366}{4500} = 4.022 \text{ m}$$

BILGING OF A MIDSHIPS SIDE COMPARTMENT

In *Figure 10.4* a box vessel has length, L; breadth, B; and is floating at draft, d_i at waterline, W_0L_0 the centre of gravity is at G and displacement W.

An empty midships side compartment: length, l; breadth, b; extending the full depth of the vessel is bilged.

The vessel sinks to water line W_1L_1 to draft d_b. The centre of buoyancy will move to B_1. If the horizontal distance between displacement acting down through G and up through B_1 is x, there will be a heeling moment $W \times x$.

If the bilged metacentre is at M_b vertically above B_1, the vessel will heel until the centre of buoyancy is at B_2 vertically beneath G and this vertical will pass through M_b provided the list angle θ is not too great.

Then if the heeling arm x, intersects the vertical through B_1 at Y

$$\tan \theta = \frac{GY}{YM_B}$$

In practice the angle θ may be quite large, however, the estimate of heel obtained will give a reasonable indication of the likely effect of bilging a side compartment.

The calculation for the vessel in *Figure 10.4* has the following stages.

(i) Calculation of bilged draft

$$LBd_1 = LBd_b - lbd_b$$

(ii) Calculation of GY.

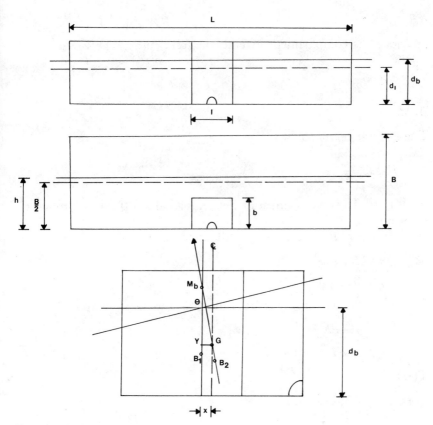

Figure 10.4 Bilging of midship side compartment

The vertical through B_1 will pass through the centroid of the water plane. In *Figure 10.4* the centroid will be a distance h from the side.

$$h = \frac{\text{moment of area about side}}{\text{area}}$$

$$= \frac{L \times B \times \dfrac{B}{2} - l \times b \times \dfrac{b}{2}}{LB - lb}$$

And since G is on the centreline a distance $B/2$ from the side:

$$GY = h - \frac{B}{2}$$

KB_1

Provided the compartment extends the full depth of the vessel

$$KB = \frac{d_b}{2}$$

$B_1 M_b$

The general relationship

$$B_1 M_b = \frac{I_b}{V}$$

holds good. However I_b must be determined about the axis through the centroid.

From *Figure 10.4* and working about the side

$$I_{side} = \frac{LB^3}{3} - \frac{lb^3}{3}$$

To find I_b we must apply the parallel axis theorem (Chapter 1)

$$I_b = I_{side} - Ah^2$$
$$= I_{side} - (LB - lb)h^2$$

Then

$$B_1 M_b = \frac{I_b}{V}$$

Then

$$KM_b = KB_1 + B_1 M_b$$
$$YM_b = KM_b - KG$$
$$\tan \theta = \frac{GY}{YM_b}$$

It is generally more convenient to work about the side rather than the centreline, however, in some cases it may be easier to work about the centreline.

Example 10.5 (*Figure 10.5*)

A box vessel length, 60 m; breadth, 9 m is floating at a draft of 5 m and has KG, 3.0 m. Find the list if empty midships side compartments length 6 m, breadth 6 m is bilged.

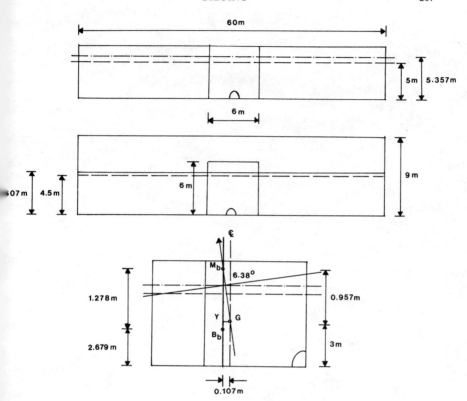

Figure 10.5 Example 10.5

Intact volume before bilging = Intact volume after bilging

$$LBd_i = LBd_b - lbd_b$$

$$60 \times 9 \times 5 = 60 \times 9 \times d_b - 6 \times 6 \times d_b$$

$$d_b = 5.357 \text{ m}$$

$$KB_1 = \frac{d_B}{2}$$

$$= \frac{5.357}{2} \text{ m} = 2.679 \text{ m}$$

h from side

	Area (m²)	Centroid (m)	Moment (m³)
$L \times B$	540	4.5	2430
$l \times b$	−36	3.0	−108
	504		2322

$$h = \frac{2322}{504} = 4.607 \text{ m}$$

$$GY = h - \frac{B}{2} = 4.607 - 4.5 = 0.107 \text{ m}$$

$$I_{\text{side}} = \frac{LB^3}{3} - \frac{lb^3}{3}$$

$$= \frac{60 \times 9^3}{3} - \frac{6 \times 6^3}{3}$$

$$= 14\,580 \text{ m}^4 - 432 \text{ m}^4 = 14\,148 \text{ m}^4$$

$$I_{\text{b}} = I_{\text{side}} - Ah^2$$

$$= 14\,148 \text{ m}^4 - 504 \times 4.607^2 \text{ m}^4$$

$$= 3451 \text{ m}^4$$

$$B_1 M_{\text{b}} = \frac{I_{\text{B}}}{\nabla}$$

$$= \frac{3451}{2700} \text{ m}$$

$$B_1 M_{\text{b}} = 1.278 \text{ m}$$

$$KB = 2.679 \text{ m}$$

$$KM_{\text{b}} = 3.957 \text{ m}$$

$$KG = 3.000 \text{ m}$$

$$YM_{\text{b}} = 0.957 \text{ m}$$

$$\tan \theta = \frac{GY}{YM_{\text{B}}}$$

$$= \frac{0.107}{0.957}$$

$$\theta = 6.38°$$

BILGING OF AN END COMPARTMENT

When dealing with this problem we will assume that the actual structure remains in place and that the compartment extends the full breadth of the vessel.

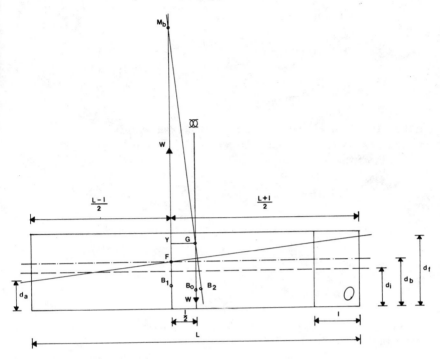

Figure 10.6 Bilging of end compartment

In *Figure 10.6* a vessel: length, L; breadth, B is floating at an even keel draft of d_1 at waterline, W_0L_0. Since the vessel is on an even keel the centre of gravity and centre of buoyancy are in a vertical line midships.

If an end compartment length l, is bilged the vessel will sink to waterline W_1L_1.

The centre of buoyancy will move to B_1 at the centroid of the intact length.

$$\text{Distance of centroid from AP} = \frac{L-l}{2}$$

The centre of gravity is $\frac{L}{2}$ from AP.

Hence the horizontal separation between

$$B_1 \text{ and } G \text{ is } \frac{L}{2} - \frac{L-l}{2} = \frac{l}{2} = GY$$

If the vessel has displacement W there is a trimming moment $W \times GY$.

The vessel will trim through

$$\text{Change of trim} = \frac{\text{moment changing trim}}{MCTC}$$

$$= \frac{W \times GY}{MCTC}$$

noting that $MCTC$ is the value for the bilged condition.

The vessel will trim about the centre of flotation F which will be at the centroid of the intact waterplane a distance $(L - l)/2$ from the **AP**.

$$\text{Change of trim aft} = \frac{L - l}{2L} t$$

The process is best illustrated with an example as follows.

Example 10.6

A box vessel has length, 180 m; breadth, 20 m is floating at an even keel draft of 12 m. KG, 8 m.

Find the drafts of the vessel fore and aft if an empty full breadth end compartment length, 12 m is bilged.

Vessel floating in fresh water.

Intact volume before bilging = Intact volume after bilging

$$L \times B \times d_i = (L - l) \times B \times d_b$$

$$180 \times 20 \times 12 = 168 \times 20 \times d_b$$

$$d_b = 12.857 \text{ m}$$

$$GY = \frac{l}{2}$$

$$= \frac{12}{2} = 6 \text{ m}$$

$$W = L \times B \times d_b \times \rho$$

$$= 180 \times 20 \times 12 \times 1.000 \text{ tonne}$$

$$= 43\,200 \text{ tonnes}$$

$$B_1 M_L = \frac{I_b}{\nabla}$$

$$= \frac{(L-l)^3 B}{12(L-l)Bd_b}$$

$$= \frac{(L-l)^2}{12d_b}$$

$$= \frac{168^2}{12 \times 12.857} = 182.9 \text{ m}$$

$$KB_1 = 6.4 \text{ m}$$

$$KM_L = 189.3 \text{ m}$$

$$KG = 8.0 \text{ m}$$

$$GM_L = 181.3 \text{ m}$$

$$MCTC = \frac{W \times GM_L}{100L}$$

$$= \frac{43\,200 \times 181.3}{100 \times 180}$$

$$= 435.1 \text{ tonne m/cm}$$

Note that L is used because in the derivation of $MCTC$, L is the length of the structure.

$$\text{Change of trim} = \frac{W \times GY}{MCTC}$$

$$= \frac{43\,200 \times 6}{435.1} = 595.7 \text{ cm}$$

$$\text{Change of trim aft} = \frac{(L-l)}{2L} \times t$$

$$= \frac{168}{2 \times 180} \times 595.7 = 278 \text{ cm}$$

Change of trim frd $= 318$ cm

	Forward (m)	Aft (m)
Bilged draft	12.86	12.86
Trim	3.18 m	− 2.78 m
Final drafts	16.04 m	10.08 m

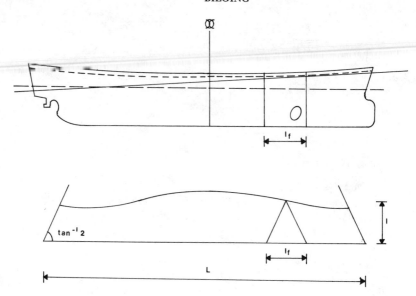

Figure 10.7 Flooding curves

The extension of these calculations to ship shapes and unsymmetrical flooding away from midships is beyond the scope of this book.

Some ships may be provided with flooding curves. These curves are primarily used for design purposes. However, under operational conditions they can be used to determine if a given amount of damage will result in the ship foundering or sinking to an unsafe condition.

The axis of the curve in *Figure 10.7* is the length of the ship. The ordinates of the curve are the length of the vessel which if flooded would result in the vessel sinking some predetermined trim and draft. There are two criteria in general use.

1. Sinking to the margin line, a line drawn 76 mm (3 ins) below the deckline at the side at the level of the bulkhead deck. This criteria is used for passenger vessels. In all cases the floodable length allowed by regulation is less than the maximum possible. This length is called the permissable length (*Figure 10.7*).
2. Sinking to the trim which would result in progressive flooding.

If the flooding curve for the vessel is *Figure 10.7* then l_f represents the length of the compartment which would just sink the vessel to the margin line.

The $\tan^{-1} 2$ lines at either end of the curve are aids to plotting.

QUESTIONS ON BILGING

1. A box vessel has length, 100 m; breadth, 10 m; and depth, 8 m. The vessel is floating at a draft of 4 m and has KG 4 m. If an empty full breadth midship compartment length 20 m is bilged, find the bilged draft and the GM before and after bilging.

2. A box vessel has length, 120 m; breadth, 14 m; is floating at a draft of 4 m. An empty midship centre line compartment length, 20 m; breadth, 6 m; depth, 3 m is bilged. Find the bilged draft and the KM bilged.

3. A box vessel has length, 100 m; breadth, 10 m; depth, 6 m is floating on an even keel at a draft of 4 m. If an empty compartment amidships has length, 20 m; breadth, 10 m with a watertight flat 4.2 m above the keel, is bilged below the flat. If the KG of the vessel is 2.8 m, find
   (a) The bilged draft;
   (b) The GM of the vessel before and after bilging.

4. A box vessel has length, 150 m; breadth, 14 m; and depth, 8 m. A midship full breadth compartment length 18 m contains coal stowage factor 1.2 m³/tonnes density 1.5 m³/tonne. If the vessel was floating at draft 5.5 m and the compartment is bilged, find the bilged draft.

5. A box vessel has length, 100 m; breadth, 8 m and is floating at draft 6 m with KG 3.2 m. A midship side compartment length, 9 m breadth, 3 m is bilged. Find the resulting list.

6. A box vessel has length, 180 m; breadth, 16 m is floating at draft, 10 m with KG 6.5 m. Find the list if a compartment midships length 10 m with longitudinal bulkheads on the centreline and 4 m to port of the centreline, is bilged.

7. A box vessel has length, 95 m; breadth, 8 m; depth, 6.5 m and is floating at draft 4 m in salt water with KG 3.0 m. Find the drafts fore and aft if a full breadth end compartment length 6 m is bilged.

BILGING (ANSWERS)

1.	Bilged draft	5.0 m
	GM intact	0.08 m
	GM bilged	0.17 m

2. Draft bilged   4.214 m
   KM   6.234 m

3. Bilged draft   4.84 m
   GM intact   1.28 m
   GM bilged   1.77 m

4. Bilged draft   5.81 m
   Permeability   0.44

5. $GT = 0.087$ m
   $TM = 0.754$ m
   List $= 6.58°$

6. $GT = 0.0282$ m
   $TM = 0.696$ m
   List $= 2.32°$ port

7. *MCTC*   50.48 tonne m/cm
   Trim   185.2 cm
   Draft   forward 5.254 m; aft 3.404 m

11 Bending moment, shear force and torsion

This chapter deals with the forces which induce bending, shearing and torsional stresses into the structure of the vessel. The actual stresses are dealt with briefly. Methods of calculating bending, shear and torsional forces in practice are considered.

OBJECTIVES

1. To define shear, bending and torsional forces.
2. To develop methods for producing buoyancy weight, load, shear force and bending moment curves.
3. To relate shear force and bending moment.
4. To consider the effects of waves on bending moment and shear force.
5. To examine the stresses induced into the vessel structure by bending moment and shear force.
6. To calculate bending moment and shear force for simple shapes.
7. To introduce methods of calculating shear force and bending moment for ship shpaes.
8. To define torsion.
9. To introduce methods of calculating torsional force in ship shapes.

SHEAR AND BENDING

So far weight and buoyancy have always been considered as being equal and opposite forces acting over the entire body of the vessel. However, if each section of the vessel is taken individually it is clear that some parts of the vessel will be heavy relative to other parts. For instance in a loaded tanker, the engine room and forepeak will be relatively light, while the cargo tanks will be relatively heavy. Therefore it is probable that weight would exceed buoyancy over the main length of the vessel, while the reverse situation would occur at the engine room and forepeak.

These changes in the relative value of weight and buoyancy give rise to shearing force and bending moment, which in turn induce shear and bending stress into the structure of the vessel. Changes in buoyancy due to waves have a similar effect.

To develop ideas of how shear and bending arise we will consider a box shaped vessel loaded at both ends having an empty midships compartment.

Shear force (*Figure 11.1*)

Taking each compartment individually:

Compartment 0 $w_0 > b_0$
Compartment 1 $w_1 < b_1$
Compartment 2 $w_0 > b_0$
$w_0 + w_1 + w_0 = b_0 + b_1 + b_0$

If it were possible to divide the vessel at each bulkhead and allow each section to float freely (*Figure 11.1*) then:

Compartment 0 $w_0 = b_{01}$
Compartment 1 $w_1 = b_{11}$
Compartment 2 $w_2 = b_{21}$

Therefore when the vessel was floating normally there must have been a force $(b_{01} - b_0)$ acting at the bulkhead between compartments 0 and 1

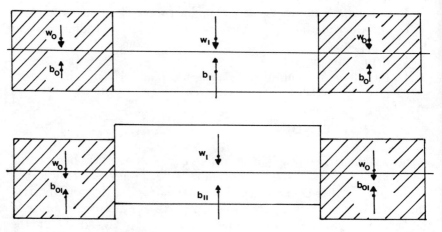

Figure 11.1 Shear force for a box shape

preventing movement from taking place

$$b_{01} - b_0 = w_0 - b_0$$

This force is the shearing force (S).

In general shearing force is equal to the difference between weight and buoyancy to the left or right of a station. There is always an equal and opposite force to the left or right of a station to balance the shearing force in order to maintain overall equilibrium.

Bending moment (*Figure 11.2*)

If the situation amidships is now considered, in this particular case weight aft of midships equals buoyancy aft of midships.

$$w_a = b_a$$

However because of the end loading the centre of gravity will be at a distance d_w aft of midships which will be greater than the distance d_b of the centre of buoyancy, aft of midships, therefore

$$w_a d_w > b_a d_b$$

There is a moment acting about midships tending to bend the vessel. Therefore there must be an equal and opposite moment acting midships to prevent the vessel bending. This is the bending moment M

$$M = w_a d_w - b_a d_w$$

In general bending moment is equal to the difference between the moment of weight and moment of buoyancy to the left or right of a particular station.

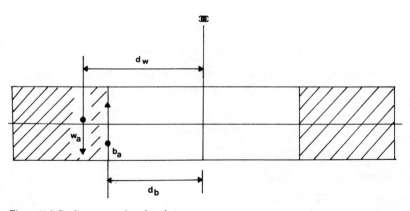

Figure 11.2 Bending moment for a box shape

It should be noted that at the bulkhead considered earlier there will be a bending moment because although the weight and buoyancy vectors are acting in the same vertical line the weight vector is greater than the buoyancy vector and hence there will be a bending moment at that bulkhead as well as a shearing force.

Determining shearing force and bending moment is therefore a matter of finding two quantities at a series of stations along the length of the vessel.

(i) The difference between weight and buoyancy to the left or right of a station.

(ii) The difference between the moment of weight and moment of buoyancy to the left or right of a station.

The systematic determination of these two values is at the core of the various methods of establishing values of shearing force and bending moment.

In order to develop ideas further we will give dimensions to the box-shaped vessel and develop curves of shear force and bending moment along the length of the vessel. *Figure 11.3(a)* the vessel has

Length, 20 m.
Length of compartments at ends, 5 m.
Light displacement, 20 tonnes (1 tonne/m).
Cargo 10 tonnes evenly distributed in *each* of the end compartments (2 tonne/m).
Load displacement, 40 tonnes (2 tonne/m).

We can now produce curves of weight and buoyancy (*Figure 11.3(b)*).

The buoyancy curve will be a horizontal line representing the uniform 2 tonne/m buoyancy. The weight curve will represent the 1 tonne/m of the light displacement, with an additional 2 tonne/m in each of the end compartments.

Area under weight curve = Area under buoyancy curve

It is more convenient to work with the resultants of the buoyancy and weight curves than to work directly with them. Therefore a load curve is produced (*Figure 11.3(c)*)

load vector = buoyancy vector − weight vector

In the case of the load curve

area above axis = area below axis

It is now possible to produce the shear force curve *Figure 11.3(d)* by taking the sum of the loads to the left of as many stations as may be needed,

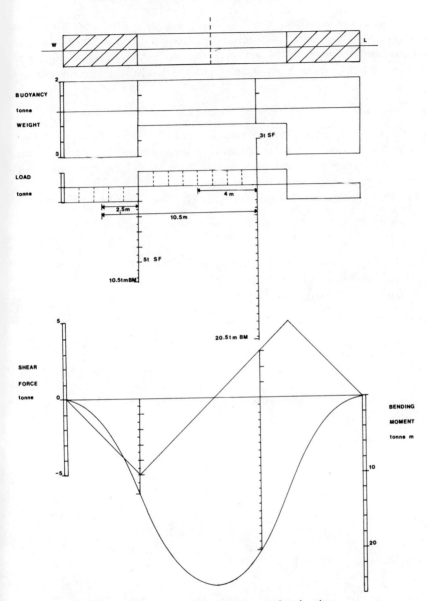

Figure 11.3 Development of shear force and bending moment curves for a box shape

i.e. 5 m from aft there are 5 tonnes of load acting to the left of the station giving a shearing force of 5 tonnes.

Again 12 m from aft there are 5 tonnes acting 'down' and 7 tonnes up giving a resultant shearing force of 2 tonnes up, giving a shearing force of

2 tonnes. The maximum shearing force occurs at the compartment bulkhead, on the system used here it is negative and the first bulkhead and position at the second, however there is no particular significance for the signs other than indicating an opposite sense of the two forces.
Again

area above axis = area below axis

Strictly, shearing force should be measured in Newtons. However, most systems for producing values of shearing force available on board ship give values in tonnes. When considering shear stress later it is important to convert to conventional force units.

Bending moment can now be found. Starting from first principles:

$$\text{Bending moment} = \text{moment of weight} - \text{moment of buoyancy}$$
$$= \text{moment of load}$$

Thereby by measuring moment of load we can find bending moment.
At the station 5 m from aft.

$$\text{Moment of load} = 5 \text{ tonne m} \times 2.5 \text{ m} = 12.5 \text{ tonne m}$$

Similarly at a station 12 m from aft

$$\text{Moment of load} = 5 \text{ tonne m} \times 9.5 \text{ m} - 7 \times 3.5 \text{ m} = 23 \text{ tonne m}$$

However the direct calculation of moment of load is rather clumsy in practice.

If we consider an elementary length of the load curve dl (*Figure 11.4*). Taking moments about C

$$M + \frac{Sdl}{2} + (S - dS)\frac{dl}{2} = M + dM$$

disregarding $\quad dS \times \dfrac{dl}{2}$

$$Sdl = dM$$

Integrating $\quad \displaystyle\int_0^1 Sdl = \int_0^1 dM$

$$\int_0^1 Sdl = M$$

But $\quad \displaystyle\int_0^1 Sdl0$ is area under shear force curve

$\therefore \quad$ Area under shear force curve up to station n
$\quad\quad = $ Bending moment at station n

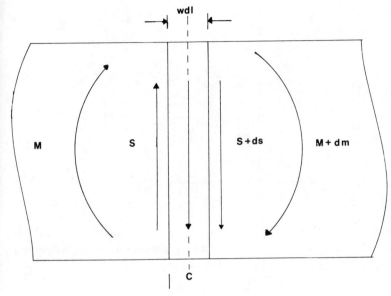

Figure 11.4 Relationship between shear force and bending moment

This is a much more convenient way in practice of establishing bending moment. The student can confirm the relationship by measuring the area under the shear force curve at stations 5 and 13 in *Figure 11.3(c)*.

The following relationships between the curves of load shear force and bending moment can be noted.

1. When the load curve crosses the zero axis shear force will be a maximum or minimum.
2. When the sum of the loads to the left or right of a station is zero, shear force will be zero.
3. When the shearing force curve crosses the zero axis bending moment will be a maximum or minimum value.
4. When the shear force is a maximum or minimum there will be a point of flexure in the bending moment curve.

In most practical conditions on ships, changes of load will occur at engine room and hold bulkheads, hence shearing force will normally reach maximum or minimum or change direction at these points. Also it is most likely that the shear force curve will cut the zero axis between bulkheads hence maximum or minimum bending moment will generally occur between bulkheads.

Therefore when calculating shearing force and bending moment it is good practice to arrange stations at bulkheads and midway between

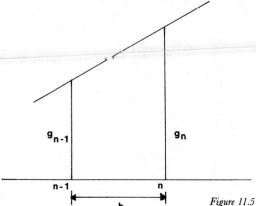

Figure 11.5 *Calculation of element of bending moment*

bulkheads (*Figure 11.6*). These stations cannot guarantee to find all the maxima and minima, especially for bending moment, but they will give a good indication of where these values occur.

Example 11.1 (*Figure 11.6*)

A box shaped vessel has length, 72 m; breadth, 12 m; and is floating at an even keel draft of 1 m in fresh water. She loads 432 tonnes of cargo into a midships compartment, length 24 m.

Produce curves of shearing force and bending moment. Find the maximum values of shear force and bending moment and the points at which they occur.

$$\text{Light displacement} = LBd\rho$$

$$= 72 \times 12 \times 1 \times 1 \text{ tonne}$$

$$= 864 \text{ tonnes}$$

$$\text{Structure per unit length} = \frac{864}{72} \text{ tonne/m}$$

$$= 12 \text{ tonne/m}$$

$$\text{Cargo per unit length} = \frac{432}{24} \text{ tonne/m}$$

$$= 18 \text{ tonne/m}$$

$$\text{Load displacement} = \text{light displacement} + \text{cargo}$$
$$= (864 + 432) \text{ tonne}$$

$$= 1296 \text{ tonne}$$

Table 11.1

Station	a Structure (tonne/m)	b Cargo (tonne/m)	c Weight (tonne/m)	d Buoyancy (tonne/m)	e Load (tonne/m)	f ΔSF (tonne)	g SF (tonne)	j ΔBM (tonne/m)	k BM (tonne/m)
0	12						0		0
1	12	0	12	18	6	36	36	108	108
2	12	0	12	18	6	36	72	324	432
3	12	0	12	18	6	36	108	540	972
4	12	18	30	18	−12	36	144	756	1728
5	12	18	30	18	−12	−72	72	648	2376
6	12	18	30	18	−12	−72	0	216	2592
7	12	18	30	18	−12	−72	−72	−216	2376
8	12	0	12	18	6	−72	−144	−648	1728
9	12	0	12	18	6	36	−108	−756	972
10	12	0	12	18	6	36	−72	−540	432
11	12	0	12	18	6	36	−36	−324	108
12	12						0	−108	0

284

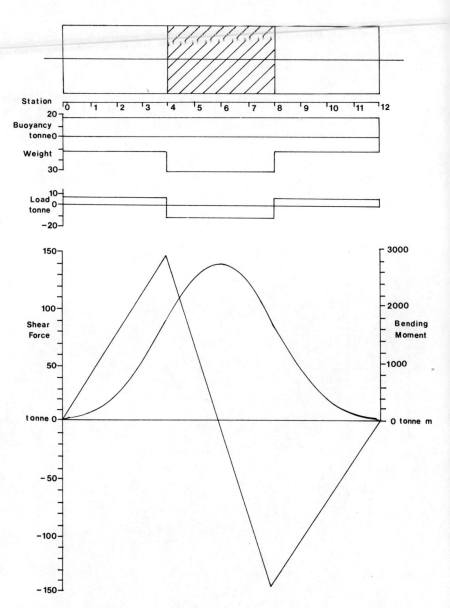

Figure 11.6 Example 11.1

$$\text{Buoyancy per unit length} = \frac{1296}{72} \text{ tonne/m}$$

$$= 18 \text{ tonne/m}$$

In order to give a clear indication of the form of the bending moment curve stations will be taken at 6 m intervals. Note that all calculations are carried out between stations. Only the values of shear force and bending moment are found at the station.

It is not essential to take stations at equal intervals. However, mistakes are less likely if stations are taken at equal intervals.

The columns of *Table 11.1* have the following values:

a = structure per unit length.
b = cargo per unit length.
$c = a + b$
d = buoyancy per unit length.
$e = d - c$

Expressed as $(d - c)$ to be consistent with the initial description there is no reason why it should not be written as $e = c - d$.

If the interval between stations is h

$$f = eh$$

$$g_1 = f_{01}$$

$$g_2 = g_1 + f_{12}$$

$$g_n = g_{n-1} + f_{(n-1)n}$$

$$j_{(n-1)n} = \frac{h}{2}(g(n-1+g_n) \quad (\text{see Figure 11.5})$$

$$k_1 = j_{01}$$

$$k_2 = k_1 + j_{112}$$

$$k_n = k_{(n-1)} + j_{(n-1)n}$$

The basic principles established for a box shape can be extended to ship shapes.

The weight curve can be determined for the light ship and then supplemented by the weight of cargo fuel etc.

The loaded drafts can be established using the standard methods for determining the vessels loaded condition. Buoyancy can then be established using Bonjean curves. For a typical bulk carrier in load condition curves of shear force and bending moment could be developed as shown in *Figure 11.7*.

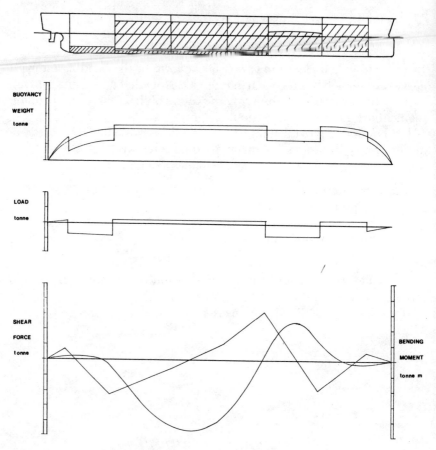

Figure 11.7 Shear force and bending moment for ship in still water

EFFECT OF WAVES

So far only the condition of the vessel in still water has been considered, however, the vessel will generally have to operate in waves. Vessels will normally have two limiting values for shearing force and bending moment, one, higher value, for still water, to cover the loading and discharging phase of operations, the other lower value to cover the vessel in her seagoing condition. The lower value is necessary to ensure that the vessel is not subjected to excessive shearing force and bending moment while at sea.

The vessel is assumed to be balanced on a trochoidal wave having length L and height $0.607\sqrt{L}$, when L is in metres. A troichoid is the

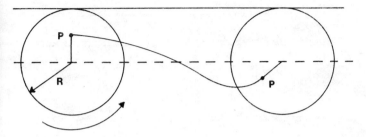

Figure 11.8 Trochoid wave form

curve produced by a point P within a circle radius R when the circle rolls along a flat base (*Figure 11.8*).

We are not concerned with the mathematics of the trochoid or the details of balancing the vessel on the trochoidal wave.

The vessel is balanced on a hogging and sagging trochoidal wave in a series of conditions and the effect on shearing force and bending moment is calculated. The effect of a sagging wave on the vessel used in *Figure 11.7* are shown in *Figure 11.9*.

BENDING STRESS AND SHEAR STRESS

At this point it is important to emphasise that the failure or otherwise of the ship structure will be caused by shear stress, or more probably bending stress, induced in the structure by shear force and bending moment. The magnitude of these forces depends upon the values of shearing force and bending moment and on the amount and distribution of material in the structure of the vessel. Since the operator has no control over the distribution or amount of material in the vessel, this book will only deal qualitatively with these stresses and how modification of the structure is likely to affect this distribution. A full account is given in *Naval Architecture for Marine Engineers* (Muckle) published by Butterworths.

Bending stress (*Figure 11.10*)

If the continuous longitudinal material has second moment of area I about the neutral axis NA and is subjected to a bending moment M, then material a distance ym from the neutral axis will have a bending stress σ.

$$\sigma = \frac{M}{I} y$$

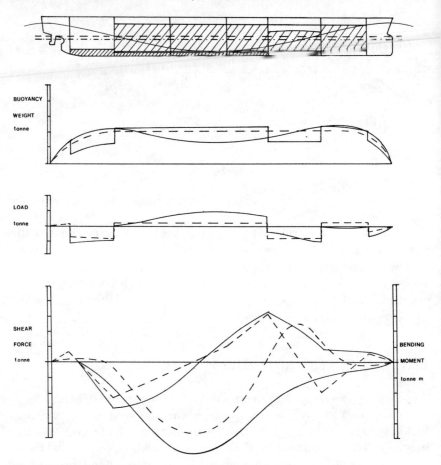

Figure 11.9 Shear force and bending moment for a ship in a sagging wave

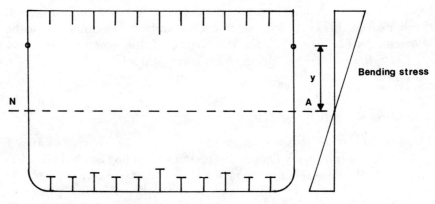

Figure 11.10 Distribution of bending stress

Since for any given condition M and I are constant the amount of stress in the structure will vary directly with the distance y from the axis NA. Then the stress will be maximum at the deck and bottom and distributed as shown in the figure. Any increase in M will increase the bending stress. Any reduction in I will increase bending stress.

The original value of I is in the hands of the designer. During the lifetime of the vessel I may be reduced by corrosion. Damage to longitudinal material can effectively reduce I. Adding superstructure can shift the neutral axis and result in change in I and the magnitude and distribution of stress.

Shearing stress (*Figure 11.11*)

If the continuous longitudinal material has second moment of area I about neutral axis NA. Then if the shear stress a distance y from NA is to be found and

$b = \Delta b_0 + \Delta b_1$

  $=$ breadth of material at y

$A =$ area of material above y

$\bar{Y} =$ centroid of area A above y

$S =$ shear force on section

$\tau =$ shear stress

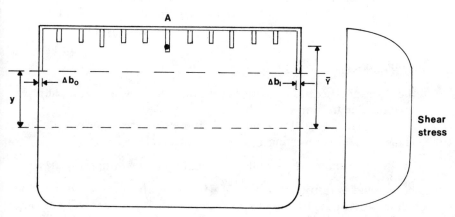

Figure 11.11 Distribution of shear stress

Then

$$\tau = \frac{SA\bar{Y}}{Ib}$$

Ib will be approximately constant for the side of the vessel. S is a constant for a given section in a given loaded condition. Area A will reduce as stress is found further from the axis NA. $\bar{Y}$ will increase slowly as the distance from the neutral axis is increased. The net result is that shear stress is a maximum at the neutral axis and reduces steadily towards the deck and bottom plating.

Maximum shear stress will typically be about one-sixth of the maximum bending stress.

Shear stress failure would appear as a wrinkling of plates near bulkheads near the neutral axis. For a typical vessel loaded homogeniously the effect of stress can be summarised as:

	Maximum	*Magnitude*	*Appearance*
Bending stress	Deck and bottom	K	Fracture of deck plating
Shear stress	Neutral axis	$\dfrac{K}{6}$	Wrinkling of plates

CALCULATION OF SHEAR FORCE AND BENDING MOMENT FOR SHIPS

Single point bending moment

This method finds the bending moment at a single point. *Table 11.2* and *Figure 11.12* illustrates the method used to find bending moment at frame 121, probably just forward of the accommodation on the vessel. The table is used to find moments for trim as well as bending moment. The correction term:

Total weight × 90.13

is the $w \times d$ term necessary to transfer the moment of weight from the after perpendicular to frame 121.

The displacement 59 500 tonnes is from elsewhere in the calculation.

This method is reverting to the basic principle that:

Bending moment = moment of weight − moment of buoyancy

The method is restricted to vessels where the operator can usually be confident that the position of maximum bending moment will not vary significantly in the loaded condition.

Table 11.2 Moments of weight about after perpendicular

Bay	Tonnes	L.C.G.	Long. Moments
\	\	Cargo Loaded This Port	
01		225.82	
02		222.75	
03	11	219.32	2 413
05	40	211.02	8 441
06		207.95	
07		204.52	
09	192	196.44	37 716
10	76	193.37	14 696
11	204	189.94	38 748
13	100	180.98	18 098
14		177.91	
15	392	174.48	68 396
17	341	164.64	56 142
18		161.57	
19	368	158.14	58 196
21	343	149.18	51 169
22		146.11	
23	382	142.68	54 501
25		132.60	
26		131.53	
27		128.10	
29	449	119.14	53 494
30		116.07	
31	440	112.64	49 562
33	778	102.67	79 877
34		99.61	
35	1926	96.18	· 185 243
	6042	Forward Fr. 121	776 694
37	18	60.77	1 094
38		57.73	
39	7	54.35	380
41		46.08	
42	48	43.00	2 064
43		39.58	
45	263	30.58	8 043
46		27.50	
47	231	24.06	
49		14.33	5 562
50		11.26	
51		7.84	
Loaded this Port	6609		793 837
Loaded previous Port	22205		2 854 987
Total	28814		3 648 824

Tabular method

If the cargo is to be loaded in a series of configurations, i.e. alternate holds empty for ore, topsided wing tanks used for ballast and grain, then it is possible that maximum bending moment and shear force will occur at

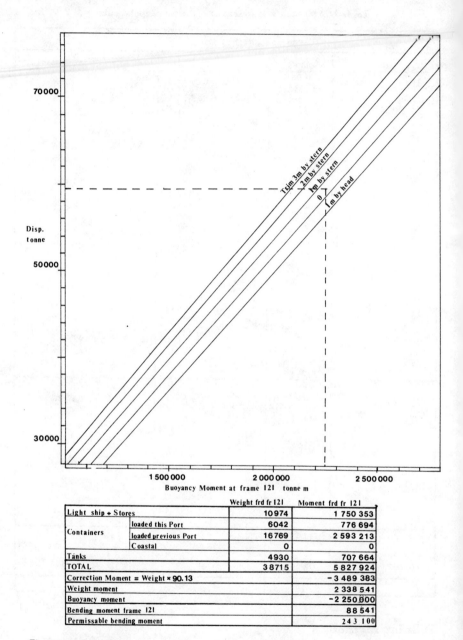

Figure 11.12 Determination of bending moment a single point (Frame 121)

		Weight frd fr 121	Moment frd fr 121
Light ship + Stores		10974	1 750 353
Containers	loaded this Port	6042	776 694
	loaded previous Port	16769	2 593 213
	Coastal	0	0
Tanks		4930	707 664
TOTAL		38715	5 827 924
Correction Moment = Weight × 90.13			– 3 489 383
Weight moment			2 338 541
Buoyancy moment			– 2 250 000
Bending moment frame 121			88 541
Permissable bending moment			243 100

several points along the length of the vessel, and that maximum bending moment will occur away from midships. Therefore there has to be a method for finding bending moment and shear force at several points along the length of the vessel.

One method is to assess the effect of loading a standard amount of cargo, typically 1000 tonnes in each on compartment at a series of stations along the length of the vessel. A section of a typical table is shown in *Table 11.3*.

The stations for shear force correspond to bulkheads for shear force. The stations for bending moment correspond to the centroid of compartments. The table is for use near specific displacement. Typically the table would be available for ballast condition and load displacement. If the tabulated values are used at displacements which are very different from the base values or are used for very uneven cargo distributions which would result in extreme trim there could be considerable error. The method is arithmetically laborious and is generally applied within a loading computer.

In *Table 11.3* weights are given in units of 1000 tonnes. Thus loading 1000 tonnes in No 1 hold reduces the draft aft by 0.408 m, increases the draft forward by 0.954 m, increases the bending moment at frame 53 by 6030 tonne m, increases the shear force at frame 37 by 193 tonnes m and so on.

Similarly loading 2400 tonnes in No 2 hold produces the changes shown in the second row of the table. Light ship values are given as data. In this method the shear force has to be corrected as shown in the lower box.

QUESTIONS ON BENDING MOMENT AND SHEARING FORCE

1. A box shaped vessel has length 80 m and light displacement 800 tonnes.

    The vessel loads 200 tonnes in a forward compartment length, 20 m and 200 tonnes in an after compartment length, 20 m. Produce curves of shearing force and bending moment.

2. A box shaped vessel has length 120 m and light displacement 1440 tonnes. Cargo is loaded into six compartments of equal length as follows:

Table 11.3 Determination of bending moment and shear force using tabular method

Compartment	Weight tonne	Draft m At A.P.	Draft m At F.P.	Bending Moment (1000 tonne m) Fr 53	Fr 86	Fr 115	Fr 143	Fr 174	Fr 37	Fr 70	Shearing Force (1000 tonne) Fr 103	Fr 127	Fr 160	Fr 188
No1 hold	1.0	−.408 −.408	.954 .954	6.03 6.03	11.65 11.65	13.77 13.77	10.39 10.39	.50 .50	.193 .193	.219 .219	.080 .080	.125 .12	.548 .548	.065 .065
No2 hold	2.4	−.163 −.391	.698 1.675	1.77 4.25	1.78 4.27	−1.48 −3.55	−8.48 −15.55	−1.37 −3.29	.055 1.32	.003 .007	−.152 −.365	−.331 −.494	.336 .806	.047 .113
No3 hold	—	.082	.472	−2.48	8.08	—	—	—	—	—	—	—	—	—
No4 hold	—	—	—	—	—	—	—	—	—	—	—	—	—	—
No5 hold	—	—	—	—	—	—	—	—	—	—	—	—	—	—
No1 TSWT	—	—	—	—	—	—	—	—	—	—	—	—	—	—
No2 TSWT	—	—	—	—	—	—	—	—	—	—	—	—	—	—
No3 TSWT	—	—	—	—	—	—	—	—	—	—	—	—	—	—
No4 TSWT	—	—	—	—	—	—	—	—	—	—	—	—	—	—
No5 TSWT	—	.682	−.132	−8.41	−3.02	—	—	—	—	—	—	—	—	—
Light ship	—													
TOTAL a	7,462	6.444	−.853	68.8	82.6	64.2	34.6	6.9	2.31	.46	−.90	−1.33	−1.14	−.31

Maximum value

Load 61.3   Ballast 78.1   Sheltered water 109.0

c diff b	b (absolute value)	3.15	3.16	3.64	3.38	3.07	3.22
d Multiplier		.142	.142	.197	.142	.137
e c × d						
f Shear force a − f						

No.	Cargo (tonnes)
1	500
2	0.0
3	400
4	400
5	0.0
6	500

Produce curves of shearing force and bending moment.

3. A box shaped vessel has length, 120 m and light displacement, 1440 tonnes. She loads cargo in six compartments of equal length as follows:

No.	Cargo (tonnes)
1	300
2	0
3	300
4	200
5	200
6	200

Produce curves of shear force and bending moment.

4. A box shaped vessel has length, 100 m and light displacement, 1000 tonnes. 500 tonnes of cargo is loaded into amidships compartment length, 20 m. Produce curves of bending moment and shearing force.

5. A box shaped vessel has length, 140 m; light displacement, 5600 tonnes, loads cargo as follows in a compartment of equal length.

No.	Cargo
1	400
2	1000
3	0
4	1000
5	0
6	1000
7	400

Produce curves of shearing force and bending moment.

BENDING MOMENT AND SHEARING FORCE (ANSWERS)

1. Length	0	10	20	30	40	50	60	70	80 m
SF	0.0	50.0	100.0	50.0	0.0	−50.0	−100.0	−50.0	0.0 tonnes
BM	0.0	250.0	1000.0	1750.0	2000.0	1750.0	1000.0	250.0	0.0 tonnes

2.
Length	0	20	40	60	80	100	120	m
SF	0.0	200.0	−100.0	0.0	100.0	−200.0	0.0	tonne
BM	0.0	2000.0	3000.0	2000.0	3000.0	2000.0	0.0	tonne m

Max. BM 3320 tonne m at 33.2 m and 86.7 m from aft.

3.
Length	0	20	40	60	80	100	120
SF	0.0	100.0	−100.0	0.0	0.0	0.0	0.0
BM	0.0	1000.0	1000.0	0.0			

Max. BM 1550 tonnes 30 m from forward.

4.
Length	0	20	40	60	80	100	m
SF	0.0	−100.0	−200.0	2000.0	100.0	0.0	tonne
BM	0.0	−1000.0	−4000.0	−1000.0	−1000.0	0.0	tonne m

Max. BM 5000.0 tonne m midships.

5.
Length	0	20	40	60	80	100	120	140	
SF	0	−143	314	−229	229	229	143	0.0	

Length	0	10	30	50	70	90	110	130	140
BM	0	−357	−1714	2072	0.0	2072	−1714	−357	0.0

TORSION

Torsion is the term used to describe the effects on a structure when it is subjected to torque. Torque can be regarded as the sum of the turning moments acting about the longitudinal axis of a structure. If a body subjected to torque is not free to rotate then the structure will twist and stresses will be induced in the structure. A typical structure subjected to torsion is the engine, propeller shaft propeller system.

For most ships torsion is not induced by normal cargo operations, although all ships are subject to torsion in oblique sea waves. However container vessels can be subjected to torsion due to containers being loaded so that:

(a) the vessel is upright or very nearly upright.

Moments port ≃ moments starboard.

(b) Containers are loaded so that a moment to port in bay 01 is balanced by a moment to starboard in say bay 14.

In this way there will be a series of torques applied along the length of the vessel (*Figure 11.13*)

$$w_0 d_0, w_1 d_1 \ldots w_1 d_1 \ldots w_d d_n$$

At any point along the length of the vessel the torque (torsional moment) will be the difference between the net moment forward of the station and proportion of the total torsional moment acting on the vessel at that

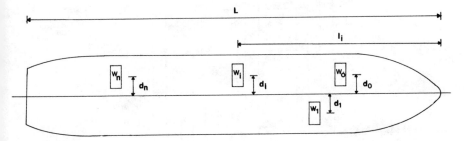

Figure 11.13 Torsional moments in a ship

station, i.e. if at each station along the length of the vessel there is a moment $M_1 = w_1 d_i$ and a total moment H along the length of the vessel.

$w_i d_i$ = moments port + moments starboard,

where moments to port are + and moments starboard −

$$H = M_1 + M_2 + \ldots M_i + \ldots M_n$$

and if station M_i is a distance l_i from forward in a vessel length L, torsional moment at station i is

$$T_i = (M_1 + M_2 \ldots M_i) - \frac{l_i}{L} H$$

Note that it is quite possible for

$$H = 0$$

and that if $H60$ the vessel will be listed.

Similarly for any particular station it is possible to have

$$M_1 + M_2 \ldots + M_i = 0$$

The data on board ship will give a proforma for calculating T_i and also a maximum value for T_i at each station.

A section of a typical proforma is given in *Table 11.2*. This particular proforma is used on the same vessel as uses the tables presented in *Tables 11.4* and *Example 11.2*

Example 11.2

A container vessel loads containers in bay as follows:

Row No	10	08	06	04	02	01	03	05	07	09
Dist. ℓ	12.28	9.62	6.95	4.00	1.34	1.34	4.00	6.95	9.62	12.28
Weight	10.62	14.86	8.62	11.00	8.00	7.36	14.62	12.28	7.62	15.27
			P					S		

Table 11.2 Table for determining torsional moment at a station

OD Row	12	10	08	06	04	02	00	01	0	03	05	07	11
21 D	14.77X	12.31X	9.85x	7.39x	4.92x	2.45x	⊠	2.45x	4.92x	7.39x	9.85x	12.31x	14.77x
21 H		12.28x	9.62x	6.95x	4.00x	1.34x		1.34x	4.00x	6.95x	9.62x	⊠	
22 D	14.77x	12.31x	9.85x	7.39x	4.92x	2.45x	⊠	2.45x	4.92x	7.39x	9.85x	12.31x	14.77x
23 D	14.77x	12.31x	9.85x	7.39x	4.92x	2.45x	⊠	2.45x	4.92x	7.39x	9.85x	12.31x	14.77x
23 H		12.28x	9.62x	6.95x	4.00x	1.34 X	1.34x	4.00x	6.95x	9.62x	12.28x		

Total brought forward

Carried forward

Bay No / Hold Row

	10	08	06	04	02	01	03	05	07	09

Torsion B

Exports						Total	
Loaded this port		Loaded prev. port		Total on departure		Total	
+P	−S	+P	−S	+P	−S	+P	−S

The remaining bays at frame 94 have a transverse moment of $+123$ tonne m. At the remaining torsional moment stations the torsional moments are as follows:

Frame No	114	104	94	84	72	40	AP
M_i	-862	-981		$+102$	-1041	$+1982$	$+219$
l	39.3	67.6	96.3	125.1	154.4	183.1	216
Max. T_M	1310	2250	3200	3030	2050	1090	0

$L = 216$.

Find the torsional moment at each station.

Does the torsional moment exceed the maximum permissible value?

+Port				−Starboard			
Row No	Dist.	Weight	Moment	Row No	Dist.	Weight	Moment
10	12.28	10.62	130.41	01	1.34	7.36	9.86
08	9.62	14.86	142.95	03	4.00	14.62	58.48
06	6.95	8.62	59.91	05	6.95	12.28	85.48
04	4.00	11.00	44.00	07	9.62	7.62	73.30
02	1.34	8.00	10.72	08	12.28		187.57
			$+387.99$				-414.69
							387.99
							-26.70
							$+123.0$
						Remaining moments	96.30

Torsional moment calculation

Frame No	Transverse moment M_i	Sum of transverse $\sum M_i$	$\dfrac{l}{L}$	$\dfrac{lH}{L}$	Torsional moment $\sum M_i - \dfrac{lH}{L}$	Max. value
114	-862	-862	0.182	-70.8	-791.2	1310
104	-981	-1843	0.313	-121.8	-1721.2	2250
94	$+192$	-1651	0.446	-173.5	-1477.5	3200
84	$+102$	-1549	0.579	-225.2	-1323.8	3030
72	-1046	-2590	0.715	-278.1	-2311.9	2050
40	$+1982$	-608	0.848	-329.9	-278.1	1090
AP	$+219$	-389	1.000			

$H = -389$.

Torsional moments are exceeded at frame 72.

12 Squat interaction and turning

This chapter is intended to give a qualitative introduction to the forces which cause vessels to squat in shallow water and interact as they pass each other. The process of turning is briefly considered.

OBJECTIVES

1. To describe the process of squatting.
2. To describe methods of predicting squat.
3. To describe the processes of interaction.
4. To describe the process of turning.
5. To consider factors affecting turning.

SQUAT

Squat is the term used to describe the changes in draft and trim which occur when the depth of water beneath the vessel is less than one-and-a-half times the draft and the vessel is travelling at considerable speed. It may also be necessary to consider squat when reading drafts when there is little under keel clearance on a berth with a considerable tide running.

The initial stages of squat can be described using Bernoulli's Theorem. This theorem deals with the conservation of energy in a liquid.

The theorem states that for any liquid:

Potential energy + pressure energy + kinetic energy = constant

$$mgZ + \frac{mp}{\rho} + \frac{1}{2}mv^2 = \text{constant}$$

Assuming liquid does not change height and therefore the potential energy term remains constant

$$\frac{mp}{\rho} + \frac{1}{2}mv^2 = \text{constant}$$

thus if the velocity of the liquid is increased then the pressure energy term must be reduced. If this happens beneath a ship the reduction in pressure must result in sinkage and because the centre of buoyancy will move due to changes in distribution of pressure the ship will also change trim.

In *Figure 12.1(a)* a vessel travelling at speed v_s in deep water does not significantly disturb the main body of water and the ship continues at her underway draft and trim. In *Figure 12.1(b)* the vessel enters shallow water, her passage will force the water to accelerate to v_w, this will in turn reduce the buoyancy vector to B_0. The vessel will sink to restore equilibrium (*Figure 12.1(c)*) and the buoyancy vector will move to B_1. There will be a trimming moment and in the figure the vessel trim by the stern (*Figure 12.1(d)*).

At this point the flow of water past the vessel is becoming more complex, the ship will have slowed down due to the increased resistance following the changes in sinkage and trim. The simple Bernoulli flow will not give a satisfactory assessment of squat.

There has been considerable research aimed at giving the mariner a method of assessing the sinkage and change in trim due to squat. *Figure 12.2* illustrates a method developed at the National Maritime Institute.[1] Using the curves and following the dashed line to take into account, water depth, trim and ship length an estimate of the sinkage and trim for a full formed tanker up to 300 m length can be made.

This method is considered to give reasonable results for ratios of depth of water to draft in the range 1.1 to 1.5. The method was developed using models; however, it has been found to give good predictions when compared to a well documented grounding due to squat.[2]

An alternative method[3] gives squat in terms of:

$$S = \text{blockage factor} = \frac{\text{midship } CSA}{\text{channel } CSA}$$

and block coefficient C_B

$$\text{Squat} = \frac{C_B S^{2/3} V_K^{2.08}}{33}$$

or in simplified form as

$$\text{Squat} = \frac{C_B V_K^2}{100} \text{ in open water}$$

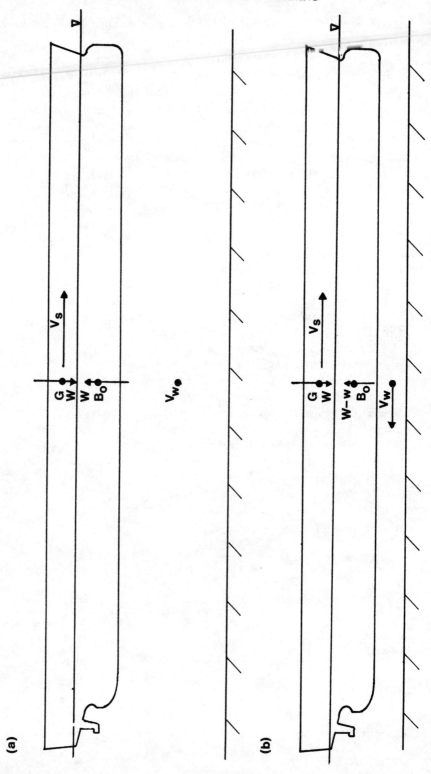

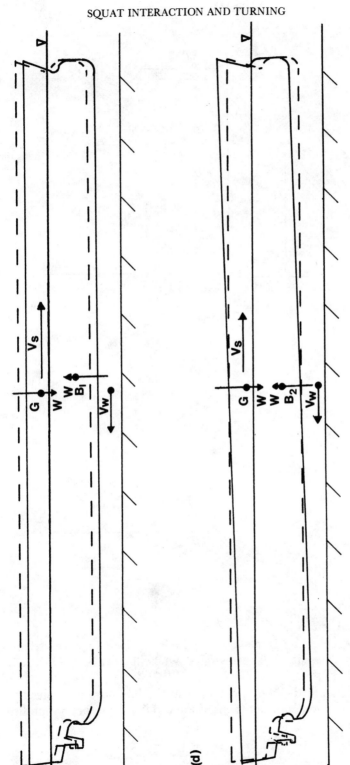

Figure 12.1 Squat in shallow water

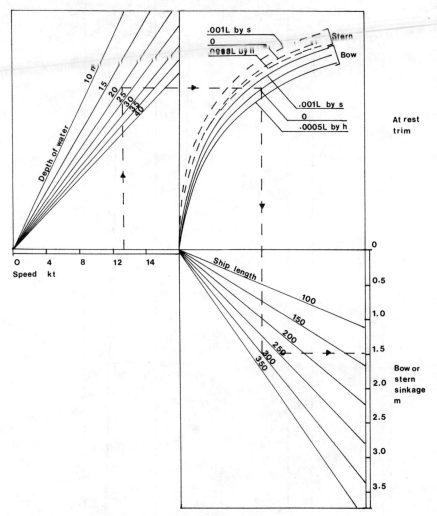

Figure 12.2 Nomogram for determining squat for full formed tanker

or

$$\text{Squat} = \frac{2C_\text{B}V_\text{K}^2}{100} \text{ in confined water}$$

There is general agreement that there is a velocity squared term when considering squat. Thus slowing down will always produce a rapid reduction of squat.

Discussion of the two approaches in the literature indicates that the graphical approach is more conservative in that it predicts grounding due

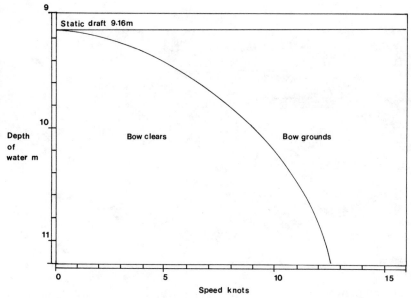

Figure 12.3 Single condition squat for bulk carrier

to squat at a lower speed than does the method using the formula. It is not for the author to make specific recommendations on the conflicting claims. Extensive discussion of the methods is available in references.

The graphical method is not available for all types of vessel. On the other hand the formula depending upon block coefficient and speed can be applied generally, although there are considerable doubts about the assumptions made to achieve this general application.

There is no doubt that the prudent mariner should apply the well tried practice of slowing down when approaching shallow water and if unexpected squat is encountered, reducing engine revolutions as rapidly as is possible and consistent with maintaining steerage.

Figure 12.3 is an example of data as supplied to a ship for assessment of squat in single condition. The curve is from the same area of research that produced the graphical approach but is taken from the stability and handling data supplied to a bulk carrier.

INTERACTION

Interaction has a similar theoretical basis to that applied to squat. Water in a narrow channel being accelerated between two vessels passing along a channel or between the vessel and the bank of the channel. Pressure is

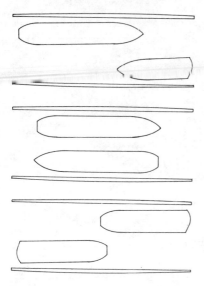

Figure 12.4 Ships passing head to head in channel

reduced between the vessels or between the vessel and a bank. The vessels move bodily towards each other or the bank and redistributed pressure over the hull causes yaw.

When interaction is being considered the simple theory is complicated by the pressure waves at the bow and stern, the effect of the screw race and the action of the rudder. Typical patterns of movement associated with overtaking, passing and bank effect are given in *Figures 12.4* and *12.5*. As with squat the most prudent course is to proceed along narrow channels at a moderate speed.

Under adverse conditions, particularly when there is a large discrepancy between the sizes of the vessels involved, it is possible that there could be a collision. Also when small vessels, tugs, etc. are close to large vessels travelling at high speed, the changes in pressure between the vessels and beneath the smaller vessel can result in the smaller vessel listing towards the larger vessel.

If the small vessel then collides with the larger vessel in extreme conditions the small vessel could capsize. This capsizing appears to be most likely when a small vessel is stationed near the bow of a larger vessel.

TURNING

Figure 12.6 gives the definitions of the terms used to describe the turning circle of a vessel. The track followed by the centre of gravity of the vessel gives the actual turning 'circle'. At the centre of gravity there will always

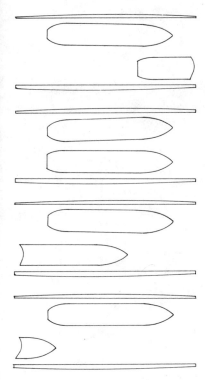

Figure 12.5 Ships overtaking in channel

be a drift angle, with the head of the vessel inboard of the circle. At some point ahead of the centre of gravity, there will be a point where the fore and aft line of the vessel is parallel to the turning circle, this is the pivot point.

When the rudder is put over a turn develops in the following stages (*Figure 12.7*):

(a) The forces of resistance F_r and propulsion F_p are in equilibrium with the rudder midships.
(b) The rudder is put over, a drag force F_d and a side force F_s develop at the rudder.
(c) The $F_r + F_d$ reduce the vessels forward speed.
(d) The F_s at the propeller can be resolved into a side force F_s acting at the centre of gravity and a moment M_s about the centre of gravity.
(e) The vessel is canted across the line of advance, with F_s and F_d again acting at the rudder, the hydrodynamic forces F_h which develop at the stern and bow can now be considered.
(f) F_h can be resolved into a side force at the centre of gravity and a moment M_h.
(g) The moments M_h and M_s are equal and opposite. The forces F_r, F_p,

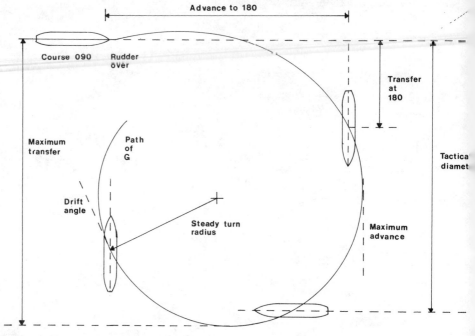

Figure 12.6 Definition of turning terms

F_s, F_h are once again considered as acting at the centre of gravity.
(h) These forces resolve into a small ahead force F_a and side force F_m.
(i) Finally these two forces resolve to F_c acting towards the centre of the turning circle and maintaining a steady rate of turn.

The size of the turning circle for a given rudder angle will depend upon forward speed, whether or not the propeller is turning and the depth of water.

Forward speed does not affect the form of the turning circle, only the time taken to complete the circle; the time taken to complete a given turn being approximately in proportion to the forward speed.

Propeller operation greatly increases rudder effectiveness. In deep water the rudder may be as much as five times more effective in terms of side force than when stopped. In shallow water the rudder may be ten times more effective when the propeller is operating than when it is stopped. Hence the value of the 'kick ahead' which can give a large increase in F_s with little change in F_r.

In deep water the side force F_h which develops as a result of the vessel canting across the flow will be small. Hence a small side force F_s at the rudder will produce a large drift angle and a high rate of turn. On the

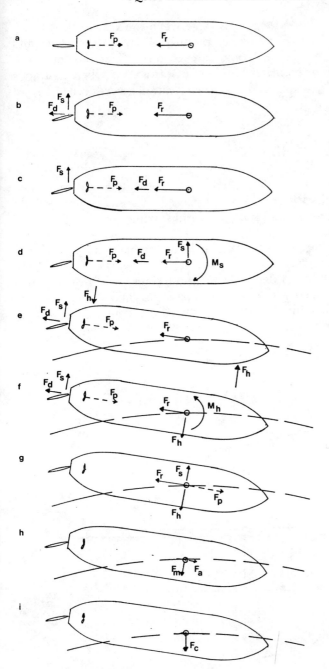

Figure 12.7 Mechanism of turning

other hand, in shallow water, F_h is large when the vessel cants, hence the rudder side force F_s will only produce a small drift angle and consequently there will be a low rate of turn. Experimental work indicates that turning diameter could increase from $3.5L$ where clearance is greater than 40% of draft to $8.74L$ when clearance is reduced to 15% of draft.

DIRECTIONAL STABILITY

A vessel is directionally stable if after putting the rudder midships while turning, the vessel settles to a steady heading. A vessel is directionally unstable if after putting the rudder midships while turning, the vessel continues to turn.

In general, fine lined ships are likely to be directionally stable while full formed ships will be directionally unstable.

Directional stability is desirable for most ships as this quality will reduce wear on steering gear on long passages, and as tugs will be available to help handling in port the reduced manoeuvrability will not be a great disadvantage.

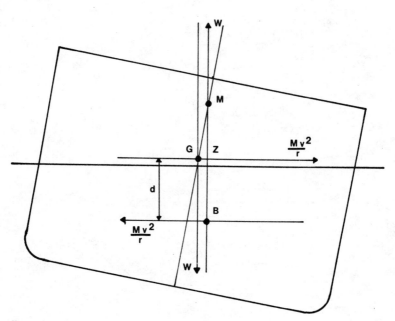

Figure 12.8 Heel due to turning

HEEL WHILE TURNING

When a vessel is in a steady turn it can be assumed that the centrifugal force is acting at centre of gravity G_1 and the equal and opposite centripedal force is acting at the centre of buoyancy B_1. The vessel will reach a steady heel when (*Figure 12.8*).

Righting moment $= d \times$ centrifugal force

where $d =$ vertical spearation between G and B_1.

$$\text{Centrifugal force} = \frac{Mv^2}{r}$$

$M =$ Mass of vessel Kg
$V =$ Velocity in m/s
$r =$ radius of turn m

Putting displacement $= Mg$

$g =$ acceleration due to gravity

$$Mg \times GZ = \frac{Mv^2}{r} \times d$$

$$GZ = \frac{v^2 d}{rg}$$

For small angles

$$GZ = GM \sin \theta$$

$$d = B_0 G \cos \theta$$

$$GM \sin \theta - \frac{v^2 B_0 G \cos \theta}{rg}$$

$$\tan \theta = \frac{v^2 B_0 G}{GMrg}$$

References

1. N.P.L. Ship Division, Estimating Bow and Stern Sinkage of a Ship Underway in Shallow Water, *The Naval Architect* No. 1 (1973)
2. FERGUSON, SEVEN and MACGREGOR, Experimental Investigation of a Grounding on a Shoaling Sandbank, *The Naval Architect* page 303 (1982)
3. BARRAS, C B, Ship Squat and its Calculation, *Safety at Sea Journal* (Feb 1978)

Index